BIOLOGY
The Dynamics of Life

BioLab and MiniLab Worksheets

New York, New York Columbus, Ohio Woodland Hills, California Peoria, Illinois

A GLENCOE PROGRAM

BIOLOGY: THE DYNAMICS OF LIFE

Student Edition
Teacher Wraparound Edition
Laboratory Manual, SE and TE
Reinforcement and Study Guide, SE and TE
Content Mastery, SE and TE
Section Focus Transparencies and Masters
Reteaching Skills Transparencies and Masters
Basic Concepts Transparencies and Masters
BioLab and MiniLab Worksheets
Concept Mapping
Chapter Assessment
Critical Thinking/Problem Solving
Spanish Resources
Tech Prep Applications
Biology Projects
Computer Test Bank Software and Manual
Windows/Macintosh
Lesson Plans
Block Scheduling
Inside Story Poster Package
English/Spanish Audiocassettes
MindJogger Videoquizzes
Interactive CD-ROM
Videodisc Program
Glencoe Science Professional Series:
Exploring Environmental Issues
Performance Assessment in the Biology Classroom
Alternate Assessment in the Science Classroom
Cooperative Learning in the Science Classroom
Using the Internet in the Science Classroom

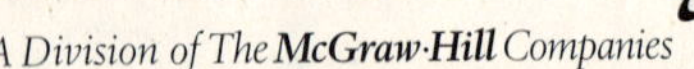

A Division of The McGraw-Hill Companies

Send all inquiries to:
Glencoe/McGraw-Hill
8787 Orion Place
Columbus, OH 43240-4027

ISBN 0-02-828256-6
Printed in the United States of America.
4 5 6 7 8 9 10 047 08 07 06 05 04 03 02 01

Contents

To the Teacher

Each activity in this booklet is an expanded version of each BioLab or MiniLab that appears in the Student Edition of ***Biology: The Dynamics of Life***. All materials lists, procedures, and questions are repeated so that students can read and complete a lab in most cases without having a textbook on the lab table. Data tables are enlarged so that students can record data in them. All lab questions are reprinted with lines on which students can write their answers. In addition, for student safety, all appropriate safety symbols and caution statements have been reproduced on these expanded pages. Answer pages for each MiniLab and BioLab are included at the end of this booklet.

Equipment and Materials List

These easy-to-use tables of equipment and consumable materials can help you prepare for your biology classes for the year. Quantities listed for BioLabs and MiniLabs are the maximum quantities you will need for one student group for the year. The Student Edition pages on which each item is used are listed in parentheses after the quantities.

Non-Consumables

Item	BioLab	MiniLab
aquarium	1 (p. 706)	
balance, laboratory	1 (pp. 780, 860)	
ball, golf	1 (p. 860)	
ball, Ping-Pong	1 (p. 860)	
beaker, 100-mL		4 (pp. 38, 155, 204, 242)
beaker, 250-mL	5 (pp. 60, 560, 706, 780, 800)	2 (pp. 23, 126, 654, 818)
beaker, 400-mL	1 (p. 168)	
beaker, 1000-mL	1 (p. 244)	
brush, camel hair	1 (p. 734)	
clock or watch with second hand	1 (pp. 26, 168, 244, 940, 1020)	1 (pp. 159, 900, 1013)
coverslip	5 (pp. 60, 194, 538, 646, 678, 800)	10 (pp. 6, 75, 88, 177, 187, 260, 379, 522, 527, 546, 606, 629, 640, 695, 703, 871)
culture dish	4 (pp. 754, 834)	1 (p. 719)
dichotomous key for trees		1 (p. 456)
dropper	5 (pp. 60, 88, 194, 538, 646, 678, 706, 754, 800, 904, 996)	4 (pp. 6, 14, 75, 260, 522, 527, 546, 609, 629, 640, 695, 703, 719, 871, 1019, 1038)
earthworm cross-section diagrams		1 (p. 750)
earthworm longitudinal diagrams		1 (p. 750)
flashlight or penlight	1 (pp. 754, 834)	
flower pots	2 (p. 280)	
forceps	1 (pp. 168, 194, 336, 512)	1 (pp. 159, 546, 606, 609, 763, 774, 797, 871)
graduated cylinders	1 (pp. 560, 780)	1 (p. 126)
hand lens	1 (pp. 280, 678, 754)	1 (pp. 333, 853, 925)
hole punch		1 (p. 406)
hot plate	1 (pp. 168, 560)	
incubator, 37°C	1 (p. 512)	
Internet access	1 (pp. 26, 130, 590, 882, 1074)	
jar, glass	3 (p. 88)	1 (p. 96)
knife, kitchen	1 (p. 168)	
lamp, 60-watt	9 (pp. 280, 834)	
light, 150-watt with reflector	1 (p. 244)	
light source, fluorescent	1 (p. 336)	
meterstick		1 (pp. 384, 980)
microscope	1 (pp. 60, 88, 194, 220, 538, 646, 678, 800, 996)	1 (pp. 6, 69, 75, 177, 187, 215, 260, 379, 506, 522, 527, 546, 606, 629, 640, 677, 695, 703, 719, 732, 788, 869, 871, 964, 1038, 1067)

Non-Consumables

continued

Item	BioLab	MiniLab
microscope slides	5 (pp. 60, 88, 194, 538, 646, 678, 734, 800, 996)	10 (pp. 6, 14, 75, 177, 187, 260, 379, 522, 527, 546, 606, 629, 640, 695, 703, 797, 871, 1019, 1038)
overhead projector		1 (p. 746)
paintbrush	2 (p. 336)	
pan, glass, shallow	1 (p. 754)	
pennies	100 (p. 394)	
petri dish	3 (pp. 26, 512, 706, 734, 800)	1 (p. 654)
petri dish, plastic	1 (p. 904)	1 (p. 890)
pH color chart		1 (p. 155)
plate		1 (p. 546)
protractor	1 (pp. 446, 860)	
razor blade, single-edged	1 (pp. 336, 646, 678, 734)	1 (pp. 159, 677)
ruler, metric	1 (pp. 26, 446, 474, 512, 538, 646, 754, 860, 966, 1020)	1 (pp. 159, 415, 490, 797, 937)
scissors	1 (pp. 308, 362, 538, 904, 1048)	
step stool	1 (p. 940)	
stereomicroscope	1 (pp. 706, 734, 754, 834, 904)	1 (pp. 577, 797)
stirring rod		2 (p. 23)
stoppers, one-hole, with glass tube	3 (p. 560)	
stopwatch	1 (p. 940)	
syringe with needle	1 (p. 800)	
telephone cord, coiled		1 (p. 304)
test tube	5 (pp. 560, 800)	2 (pp. 56, 654)
test tube, large	3 (pp. 560)	2 (p. 242)
test-tube rack	1 (p. 560)	
thermometer	1 (pp. 168, 280, 560, 754, 834)	1 (p. 818)
tongs	1 (p. 168)	
tray, clear plastic	1 (p. 780)	
tray, plastic		1 (p. 126)
tubing, rubber, 10-cm length	3 (p. 560)	
turkey baster	1 (p. 706)	
washers, metal	2 (p. 244)	3 (p. 242)
watch glass		1 (p. 719)
watering bottle, plant	1 (p. 280)	
weights, small	2 (p. 940)	

Consumables

Item	BioLab	MiniLab
adding machine tape		1 (p. 384)
aged tap water	5 mL (p. 996)	

Consumables

continued

Item	BioLab	MiniLab
aluminum foil, 30 cm × 30 cm	1 (p. 560)	
ammonia solution, household		5 mL (p. 155)
antacid tablet		1 (p. 56)
antibiotic disks (3 types)	3 (p. 512)	
apples, fresh		4 (p. 1060)
apples, rotting		1 (p. 1060)
bag, paper, small	1 (pp. 422, 108)	1 (p. 554)
bag, plastic, zipper-lock		8 (pp. 38, 546, 677, 1060)
balloon, round	1 (p. 1020)	
banana		1 (p. 96)
bean seeds	175 (p. 108)	5 (p. 677)
beans, pinto	50 (p. 422)	
beans, white navy	50 (p. 422)	
bird seed, assorted varieties		500 g (p. 856)
bottle, plastic milk, 1-gallon		3 (p. 856)
Brassica rapa seeds	40 (p. 336)	
bread, bakery, fresh		1 (p. 546)
brine shrimp eggs	100 (p. 780)	
cardboard	30 cm × 100 cm (p. 860)	
carrot		1 (p. 654)
celery stalk		1 (p. 629)
cellophane, assorted colors	1 (p. 244)	
clay, modeling assorted colors	100 g (p. 860)	500 g (pp. 234, 274)
coffee, caffeinated	1 mL (p. 996)	
cola	1 mL (p. 996)	
colored pencils (4 colors)	1 (pp. 362, 422, 678, 966)	1 (p. 333)
conifer branches, assorted	3 (p. 618)	
conifer cones, assorted	3 (p. 618)	
conifer twigs, assorted	3 (p. 618)	
corn seeds		5 (p. 677)
cosmetics, assorted		1 (p. 14)
cotton balls		1 (p. 1060)
cotton swabs	3 (pp. 512, 754)	
cough medicine	2 mL (p. 996)	
detergent, liquid		5 mL (p. 155)
diatomaceous earth		1 g (p. 379)
dinosaur eggs, artificial, from cereal		2 (p. 23)
egg, hard-boiled	1 (p. 860)	
filter paper disk, sterile	3 (p. 512)	
food coloring, yellow		1 mL (p. 1019)
garlic clove		1 (p. 654)

Consumables

continued

Item	BioLab	MiniLab
glucose test paper strips		6 (p. 1019)
glue		5 mL (p. 456)
graph paper, sheet of	4 (pp. 394, 442, 940, 966)	1 (p. 1042)
grass seed		1 g (p. 609)
ice	400 g (pp. 168, 754)	
ice cubes	2 (p. 560)	2 (p. 818)
index cards	2 (p. 538)	1 (p. 925)
ink pad		1 (p. 925)
juice, apple		15 mL (p. 242)
juice, lemon		5 mL (p. 155)
labels	12 (pp. 88, 336, 706, 734, 780)	2 (p. 38)
lima beans		10 (p. 609)
marking pens	1 (pp. 108, 512)	
mushroom, fresh		4 (p. 554)
onion		1 (p. 187)
owl pellet		1 (p. 871)
paper cup, small		6 (p. 40)
paper towels	2 (p. 754)	8 (pp. 38, 677, 871, 890)
paper, black, sheet of	1 (p. 754)	3 (pp. 406, 890)
paper, construction (5 colors)	1 (p. 308)	
paper, heavy, sheet of	1 (p. 1048)	1 (p. 456)
paper, notebook, sheet of	4 (pp. 26, 108, 1074)	10 (pp. 6, 22, 23, 26, 96, 105, 527, 1067)
paper, tracing, sheet of	1 (p. 1048)	
paper, white, sheet of	1 (p. 362)	2 (pp. 406, 554)
peas		10 (p. 609)
pencil	1 (pp. 26, 904)	
personal hygiene products		1 (p. 14)
pH paper strip		6 (p. 155)
pipet, plastic disposable		1 (p. 242)
plastic wrap, 30 cm × 30 cm	1 (p. 88)	
pond water	200 mL (p. 88)	300 mL (p. 527)
pond water, sterile	300 mL (p. 60)	
potato	1 (p. 168)	2 (pp. 159, 654)
puzzle, paper		1 (p. 900)
rice	10 g (p. 88)	1 g (p. 609)
rubber bands	1 (p. 904)	1 (p. 96)
rye seeds		1 g (p. 609)
sample keys from guidebooks	1 (p. 474)	
sandpaper, sheet of	1 (p. 754)	
sea urchin eggs		1 mL (p. 1038)
sea urchin sperm		1 mL (p. 1038)

Consumables

continued

Item	BioLab	MiniLab
sea water	500 mL (p. 800)	
seedling flat	2 (p. 280)	
seeds		40 (p. 38)
shampoo		5 mL (p. 155)
shoebox with lid	1 (p. 394)	
skull diagrams	1 (p. 446)	
soil, lawn with grass		1 kg (p. 126)
soil, lawn without grass		1 kg (p. 126)
soil, potting	1 kg (pp. 280, 336)	
spoon, plastic	1 (p. 88)	
spring water	150 mL (pp. 734, 904)	
spring water, pasteurized	600 mL (p. 88)	
straw, drinking	1 (p. 560)	1 (p. 56)
string	1 m (pp. 244, 860, 1020)	
tape, masking	10 cm (p. 904)	
tape, transparent	10 cm (pp. 308, 362)	10 cm (p. 890)
tea	1 mL (p. 996)	
tobacco	1 g (p. 996)	
tobacco seeds	20 (p. 280)	
toothbrush bristles		10 (p. 695)
toothpicks	4 (p. 538)	4 (p. 654)
tray, potting	1 (p. 336)	
twist ties		5 (pp. 204, 274)
unshelled peanuts		30 (p. 415)
wax marking pencil	1 (pp. 560, 706, 734)	1 (pp. 14, 1019)
wax paper, 30 cm × 30 cm	1 (p. 168)	
wheat seeds		1 g (p. 522)
wire		2 m (p. 856)
wire mesh, fine, 15 cm × 15 cm		1 (p. 96)
yeast, baker's		5 g (p. 242)

Living Organisms

Item	BioLab	MiniLab
Armadillidium (pill bug)	1 (p. 26)	5 (p. 890)
bacteria cultures	1 (p. 512)	
brine shrimp culture		1 (p. 719)
Daphnia culture	1 (p. 996)	
Didinium culture	1 (p. 60)	
earthworms	2 (p. 754)	
Elodea plant	3 (p. 244)	

Living Organisms

continued

Item	BioLab	MiniLab
Equisetum		1 (p. 586)
Euglena culture	1 (p. 538)	
fern frond		1 (p. 606)
flower (complete)	1 (p. 678)	1 (p. 260)
frog eggs, fertilized	10 (p. 834)	
goldfish		1 (p. 818)
hydra culture		1 (p. 719)
lichen		4 (p. 69)
Marchanita		1 (p. 577)
marine plankton culture		1 (p. 75)
mildew sample		1 (p. 6)
mudworms (*Lumbriculus variegatus*)	1 (p. 904)	
onion plant	1 (p. 646)	
Paramecium culture	1 (pp. 60, 538)	1 (p. 522)
planarian culture	1 (p. 734)	
plant leaves		6 (pp. 456, 640)
rotifer culture		1 (p. 695)
sea urchins (male and female)	2 (p. 800)	
vinegar eel culture		1 (p. 703)
zebrafish	10 (p. 706)	

Chemicals

Item	BioLab	MiniLab
acetone		10 mL (p. 925)
agar plate, sterile nutrient	1 (p. 512)	
aged tap water	25 mL (p. 996)	
alcohol testing solution		1 mL (p. 14)
alcohol, isopropyl (2-propanol)		1 mL (p. 14)
baking soda solution, 0.25%	50 mL (p. 244)	
bromthymol blue (BTB) solution	30 mL (p. 560)	30 mL (p. 56)
ethanol, 95%	2 mL (p. 996)	5 mL (p. 1060)
glucose (corn syrup)	20 mL (p. 560)	
glycerol (glycerin)		1 mL (p. 606)
hydrogen peroxide, 3%	1 mL (p. 168)	
iodine solution		150 mL (p. 204)
iodine stain		1 mL (p. 609)
methyl cellulose solution	1 mL (p. 538)	
nail polish remover		10 mL (p. 925)
potassium chloride solution	1 mL (p. 800)	

Chemicals

continued

Item	BioLab	MiniLab
potassium permanganate solution		100 mL (p. 159)
Ringer's solution for frogs	400 mL (p. 834)	
salt solution, 5%		1 mL (p. 640)
salt solution, 10%		50 mL (p.38)
salt, table, non-iodized	30 g (p. 780)	
starch solution		100 mL (p. 204)
vinegar		5 mL (p. 155)

Preserved Specimens

Item	BioLab	MiniLab
Bacillus subtilus slides	1 (p. 194)	
bacteria slides		1 (p. 506)
blood cells, red and white slides		1 (p. 1067)
Branchiostoma californiense (*Amphioxis*)		1 (p. 797)
crayfish		1 (p. 763)
Elodea slides	1 (p. 194)	
feather, contour		1 (p. 853)
feather, down		1 (p. 853)
fish mitosis slides		1 (p. 215)
frog blood slides	1 (p. 194)	
frog, adult		1 (p. 830)
grasshopper, life stage specimens		1 (p. 774)
human tooth cross-section slides		1 (p. 869)
Lycopodium		1 (p. 586)
moth, life stage specimens		1 (p. 774)
onion root tip slides	1 (p. 220)	
pork worm larvae slides		1 (p. 732)
sea star pedicellariae slides		1 (p. 788)
shells, marine, assorted		1 (p. 746)
tadpole		1 (p. 830)
thyroid and parathyroid slides		1 (p. 964)

Name Date Class

MiniLab 1-1

Observing

Predicting Whether Mildew Is Alive

What is mildew? Is it alive? We see it "growing" on plastic shower curtains or on bathroom grout. Does it show the characteristics associated with living things?

Procedure

1 Use the data table below.

Data Table

Prediction	Life characteristics
First	none
Second	
Third	

2 Predict whether or not mildew is alive. Record your prediction in the data table under "First Prediction."

3 Obtain a sample of mildew from your teacher. Examine it for life characteristics. Make a second prediction and record it in the data table along with any observed life characteristics. **CAUTION:** ***Wash hands thoroughly after handling the mildew sample.***

4 Following your teacher's directions, prepare a wet mount of mildew for viewing under the microscope. **CAUTION:** ***Use caution when working with a microscope, microscope slides, and coverslips.***

5 Are there any life characteristics visible through the microscope that you could not see before? Make a third prediction and include any observed life characteristics.

Analysis

1. Describe any life characteristics you observed.

2. Compare your three predictions and explain how your observations may have changed them.

3. Explain the value of using scientific tools to extend your powers of observation.

Name Date Class

MiniLab 1-2 Testing for Alcohol

Experimenting

Commercials for certain over-the-counter products may not tell you that one of the ingredients is alcohol. How can you verify whether or not a certain product contains alcohol? One way is to simply rely on the information provided during a commercial. Another way is to experiment and find out for yourself.

Procedure

1 Use the data table below.

Data Table

	Color of Liquid	Alcohol present
Circle A		
Circle B		
Circle C		
Product name		
Product name		

2 Draw three circles on a glass slide. Label them A, B, and C. **CAUTION:** ***Put on safety goggles.***

3 Add one drop of water to circle A, one drop of alcohol to circle B, and one drop of alcohol-testing chemical to circles A, B, and C. **CAUTION:** ***Rinse immediately with water if testing chemical gets on skin or clothing.***

4 Wait 2–3 minutes. Note in the data table the color of each liquid and the presence or absence of alcohol.

5 Record the name of the first product to be tested.

6 Draw a circle on a clean glass slide. Add one drop of the product to the circle.

7 Add a drop of the alcohol-testing chemical to the circle. Wait 2–3 minutes. Record the color of the liquid.

8 Repeat steps 5–7 for each product to be tested. **CAUTION:** ***Wash your hands with soap and water immediately after using the alcohol-testing chemical.***

9 Complete the last column of the data table. If alcohol is present, the liquid turns green, deep green, or blue. A yellow or orange color means no alcohol is present.

MiniLab 1-2

Testing for Alcohol, *continued*

Analysis

1. Explain the purpose of using the alcohol-testing chemical with water, with a known alcohol, and by itself.

2. Which products did contain alcohol? No alcohol?

Name Date Class

MiniLab 1-3

Observing and Inferring

Hatching Dinosaurs

"Dinosaur eggs" can be found in specially marked packages of oatmeal. You will conduct an investigation to determine what causes these pretend eggs to hatch.

Procedure

1. Use the data table below.

Data Table

	Before treatment	Hot water treatment	Cold water treatment
Appearance after one minute			

2. Observe the dinosaur eggs provided and record their characteristics in your table.
3. Place an egg in each of two containers.
4. Make a hypothesis about the water temperature that will cause the eggs to hatch.
5. Pour boiling water into one container and cold water in the other. Stir for one minute. Record your observations.

Analysis

1. Infer whether heat or moisture was more important for hatching eggs.

2. Design an experiment that would test either heat or moisture as the variable. What kind of quantitative data will you gather?

3. What will be your control?

4. How many trials will you run and how many eggs will you test? If time permits, conduct your experiment.

Chapter 1

Collecting Biological Data

PREPARATION

Problem
What life characteristics can be observed in a pill bug?

Objectives
In this BioLab, you will:
- **Observe** whether life characteristics are present in a pill bug.
- **Measure** the length of a pill bug.
- **Experiment** to determine if a pill bug responds to changes in its environment.
- **Use the Internet** to collect and compare data from other students.

Materials
pill bugs, *Armadillidium*
watch or classroom clock
container, glass or plastic
pencil with dull point
ruler
computer with Internet connection

Safety Precautions
Always wear goggles in the lab.

Skill Handbook
Use the **Skill Handbook** if you need additional help with this lab.

PROCEDURE

1. Use the data table and graph outlines shown here.
2. Obtain a pill but from your teacher and place it in a small container.
3. Observe your pill bug to determine whether or not it has an orderly structure. Record your answer in the data table.
4. Using millimeters, measure and record the length of your pill bug in the data table.
5. Using your data and data from your classmates, complete the graph "Pill Bug Length: Classroom Data."
6. Go to the Glencoe Science Web Site at **www.glencoe.com/sec/science** to **post your data**.
7. *Gently* touch the underside of the pill bug with a *dull* pencil point. It may be necessary to gently flip the pill bug over with the pencil to get at its underside. **CAUTION:** ***Use care to avoid injuring the pill bug.***
8. Note its response and time, in seconds, how long the animal remains curled up. Record the time in the data table as Trial 1.
9. Repeat steps 7–8 four more times, recording each trial in the data table.
10. Calculate the average length of time your pill bug remains curled up in a ball.

Data Table

Organization and growth and development	
Orderly structure?	
Pill bug length in mm	

Response to environment	
Trial	**Time in seconds**
1	
2	
3	
4	
5	
Total	
Average time	

INTERNET BioLab

Collecting Biological Data, *continued*

11. **Post your data** on the Glencoe Science Web Site.
12. Return the pill bug to your teacher. **CAUTION:** ***Wash your hands with soap and water after working with pill bugs.***

Pill Bug Length: Classroom Data

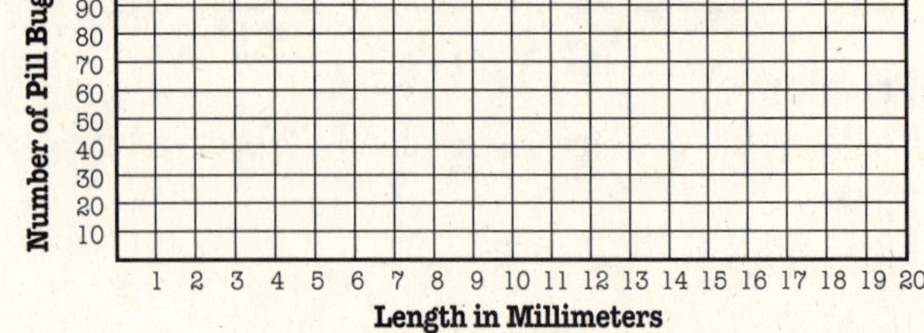

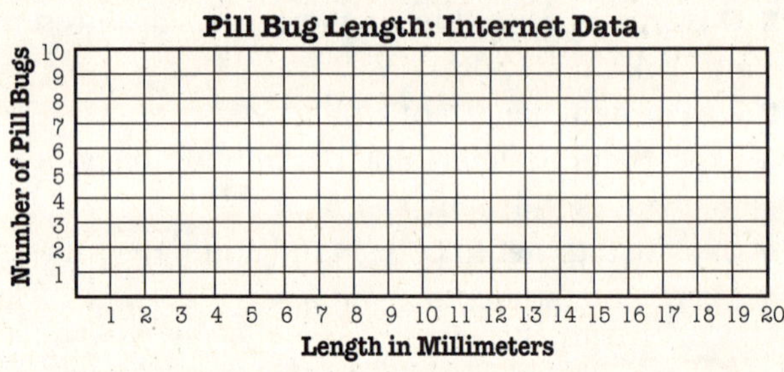

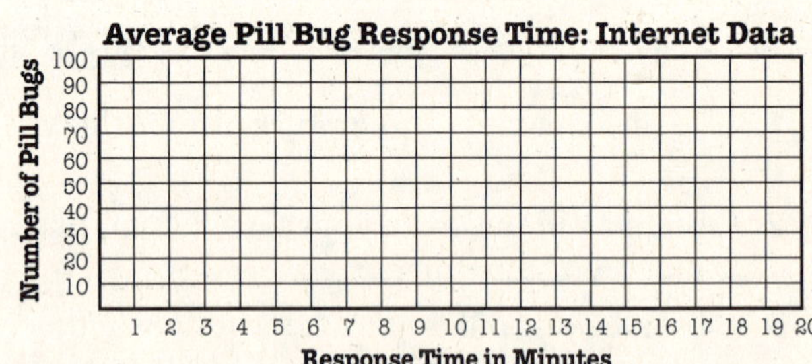

1. **Thinking Critically** Explain how you would define the term “orderly structure.” Explain how this trait might also pertain to nonliving things.

2. **Using the Internet** Explain how data from the classroom and Internet graphs support the idea that pill bugs grow and develop.

3. **Interpreting Data** What was the most common length of time pill bugs remained curled in response to being touched?

4. **Drawing a Conclusion** Explain how the response to being touched is an adaptation.

5. **Experimenting** How might you design an experiment to determine whether or not pill bugs reproduce?

Name Date Class

MiniLab 2-1

Experimenting

Salt Tolerance of Seeds

Salinity, or the amount of salt dissolved in water, is an abiotic factor. Might salt water affect how certain seeds sprout or germinate? Experiment to find out.

Procedure

1. Soak 20 seeds in freshwater and 20 seeds in salt water overnight.
2. The next day, wrap the seeds in two different moist paper towels. Slide the towels into separate self-sealing plastic bags.
3. Label the bags “fresh” and “salt.”
4. Examine all seeds two days later. Count the number of seeds in each treatment that show signs of root growth or sprouting, which is called germination. Record your data. **CAUTION:** ***Be sure to wash your hands after handling seeds.***

Analysis

1. Did the germination rates differ between treatments? If yes, how?

2. What abiotic factor was tested in this experiment? What biotic factor was influenced?

3. Might all seeds respond to salt in a similar manner? How could you find out?

Name Date Class

MiniLab 2-2

Observing and Inferring

Detecting Carbon Dioxide

Carbon dioxide is given off during respiration. When carbon dioxide is dissolved in water, an acid is formed. Certain chemicals called indicators can be used to detect acids. One indicator, called bromothymol blue, will change from its normal blue color to green or yellow if an acid is present.

Procedure

1. Half fill a test tube with bromothymol blue solution.
2. Add a quarter of an effervescent antacid tablet to the tube and note any color change.
3. Half fill a test tube with bromothymol blue solution. Using a straw, exhale into the bromothymol blue at least 30 times. **CAUTION:** ***Do not inhale the bromothymol blue.*** Record any color change in the test tube.

Analysis

1. Describe the color change that occurs when carbon dioxide is added to bromothymol blue.

____

____

2. What was the chemical composition of the bubbles seen in the tube with the antacid tablet?

____

____

3. Does exhaled air contain carbon dioxide? Explain.

____

____

Name Date Class

How can one population affect another?

Chapter 2

Preparation

Problem

How does a population of *Paramecium* react to a population of *Didinium*?

Hypotheses

Have your group agree on a hypothesis to be tested. Record your hypothesis.

Objectives

In this BioLab, you will:

- **Design** an experiment to establish the relationships between *Paramecium* and *Didinium*.
- **Use** appropriate variables, constants, and controls in experimental design.

Possible Materials

microscope
microscope slides
coverslips
culture of *Didinium*
culture of *Paramecium*
beakers or jars
eyedroppers
sterile pond water

Safety Precautions

Take care when using electrical equipment. Always use goggles in the lab. Handle slides and coverslips carefully. Dispose of broken glass in a container provided by your teacher.

Skill Handbook

Use the **Skill Handbook** if you need additional help with this lab.

Plan the Experiment

1. Review the discussion of feeding relationships in this chapter.
2. Decide which materials you will use in your investigation. Record your list.
3. Be sure that your experimental plan contains a control, tests a single variable such as population size, and allows for the collection of quantitative data.
4. Prepare a list of numbered directions. Explain how you will use each of your materials.

Check the Plan

Discuss the following points with other group members to decide final procedures. Make any needed changes to your plan.

1. What will you measure to determine the effect of the *Didinium* on *Paramecium*? If you count *Paramecia*, will you count all you can see in the field of vision of the microscope at a certain power? Will you have multiple trials? If so, how many?
2. What single factor will you vary? For example, will you put no *Didinium* in one culture of *Paramecium* and 5 mL of *Didinium* culture in another culture of *Paramecium*?
3. How long will you observe the populations?
4. How will you estimate the changes in the populations of *Paramecium* and *Didinium* during the experiment?
5. ***Your teacher must approve your plan before you proceed.***
6. Carry out your experiment.
7. Make a data table that has Date, Number of *Paramecium*, and Number of *Didinium* across the top. Place the data obtained for each culture in rows. Design and complete a graph of your data.

DESIGN YOUR OWN BioLab

Chapter 2

How can one population affect another?, *continued*

Analyze and Conclude

1. **Analyzing Data** What differences did you observe among the experimental groups? Were these differences due to the presence of *Didinium*? Explain.

2. **Drawing Conclusions** Did the *Paramecium* die out in any culture? Why or why not?

3. **Checking Your Hypothesis** Was your hypothesis supported by your data? If not, suggest a new hypothesis.

4. **Thinking Critically** List several ways that your methods may have affected the outcome of the experiment.

MiniLab 3-1 Looking at Lichens

Observing

Lichens have the reputation for being a pioneer species when it comes to succession. They often inhabit rocky areas and start the process of soil formation. How is it possible for lichens to grow on rock?

Procedure

1. Examine the lichen samples provided by your teacher. Note their color, shape, and texture.
2. Use a microscope to examine a prepared slide of a stained section of a lichen. Use low-power magnification and then change to high power as needed.
3. Observe the dark bodies that are cells containing chloroplasts. Notice that lichens are composed of an alga and a fungus. Diagram what you see.

Analysis

1. Describe the general appearance of a whole lichen and of the lichen under a microscope.

2. How does a lichen illustrate mutualism?

3. Explain how mutualism explains why lichens are able to survive on rocks.

Name Date Class

MiniLab 3-2

Comparing and Contrasting

Looking at Marine Plankton

Plankton is the term used to define the microscopic life forms present in an aquatic environment. Plankton consists mainly of protists and animal larvae.

Procedure

1. Use a medicine dropper to obtain a small sample of marine plankton.
2. Prepare a wet mount of the material. **CAUTION:** ***Handle microscope slides and coverslips carefully.***
3. Observe under low-power magnification of the microscope.
4. Look for a variety of organisms and diagram several different types.

Analysis

1. Describe the appearance of specific planktonic organisms. Draw what you see.

2. Are both autotrophs and heterotrophs present? How can you distinguish them?

3. Why are plankton important in food chains?

Name Date Class

Chapter 3

Succession in a Jar

PREPARATION

Problem

Can you observe succession in a pond water ecosystem?

Objectives

In this BioLab, you will:

- **Observe** changes in three pond water environments.
- **Count** the number of each type of organism seen.
- **Determine** if the changes observed illustrate succession.

Materials

glass jars, 3
pasteurized spring water
pond water containing plant material
labels
glass slides and cover glasses
droppers
plastic wrap
cooked white rice
teaspoon, plastic
microscope

Safety Precautions

Always wear goggles in the lab.

Skill Handbook

Use the **Skill Handbook** if you need additional help with this lab.

PROCEDURE

1. Examine the pond water sample provided by your teacher.
2. Fill three glass jars with equal amounts of pasteurized (sterilized) spring water.
3. Label them A, B, and C. Add your name and today's date to each label.
4. Add the following to each of your three jars:
   Jar A: Nothing else
   Jar B: 3 grains of cooked white rice
   Jar C: 3 grains of cooked white rice, one teaspoon of pond sediment, and a small amount of any plant material present in the pond water.
5. Record the turbidity of each jar in the data table below. Turbidity means cloudiness and can best be judged by comparing jar A to B to C. Score turbidity on a scale of 1 to 10, with 1 meaning very clear water and 10 meaning very cloudy water.
6. Gently swirl the contents of each jar.
7. Using a different clean dropper for each jar, remove a sample and prepare a wet mount of the liquid from each jar. Label each glass slide A, B, or C to avoid any mix-up. **CAUTION:** ***Handle glass slides, coverslips, and glassware carefully.***
8. Observe each slide under low power. Look for autotrophic and heterotrophic organisms. Identify these organisms by name, describe their appearance, or make a sketch of what they look like.
9. Report the number of each type of organism as viewed in a low power field of view (the circle of light seen when looking through the microscope under low power is your field of view).
10. Complete the data table for your first observations.
11. Cover each jar with either a lid or plastic wrap and place them in a lighted area.
12. Observe the jars every three days for several weeks, repeating steps 5–11 each time an observation is made. Use your data table to record your observations.

Name Date Class

Chapter 3

Succession in a Jar, *continued*

Data Table

Date	Jar	Turbidity	Name, description, or diagram of organism seen	Autotroph or heterotroph?	Number seen per low power field
	A				
	B				
	C				
	A				
	B				

Analyze and Conclude

1. **Applying Concepts** Which of the three jars was a control? Explain why.

2. **Observing and Inferring** What might have been the role of the cooked rice?

3. **Recognizing Cause and Effect** Turbidity was a means of indirectly measuring the amount of bacterial growth in the jars. Why was there little, if any, turbidity in jar A?

4. **Analyzing Information** Describe the changes over time in the number and type of heterotrophs. Could these changes be described as succession? Explain.

5. **Observing and Inferring** Were you able to observe a climax ecosystem during this experiment? Explain your answer.

MiniLab 4-1

Making and Using Tables

Fruit Fly Population Growth

Fruit Flies (*Drosophila melanogaster*) and similar insects have rapid rates of reproduction. Fruit flies are frequently used in biological research because they reproduce quickly and are easy to keep and count. In this activity, you will observe the growth of a fruit fly population as it exploits a food supply.

Procedure

1. Place half a banana in a jar and allow it to sit outside in a warm shaded area, or in a warm area in your classroom.
2. Leave the jar for one day or until you have at least three fruit flies in it. Put the mesh on top of the jar and fasten with the rubber band.
3. Each day record how many adult fruit flies are alive in the jar. Record for at least three weeks. Put your data into table form. **CAUTION:** ***Return the fruit flies to your teacher for proper disposal.***

Analysis

1. How many fruit flies did you start with? On what day were there the most fruit flies? How many were there?

2. Why did the number of fruit flies decrease?

3. Based on this investigation, why are insects considered to display a rapid reproduction pattern?

Name Date Class

Using Numbers

MiniLab 4-2 Doubling Time

The time needed for any population to double its size is known as its "doubling time." For example, if a population grows slowly, its doubling time will be long. If it is growing rapidly, its doubling time will be short.

Procedure

1. The following formula is used to calculate a population's doubling time:

$$\text{Doubling time (in years)} = \frac{70}{\text{annual percent growth rate}}$$

2. Use the data table below.
3. Complete the table by calculating the doubling time of human populations for the listed geographic regions.

Data Table

Geographic region	Annual percent growth rate	Doubling time
Africa	2.8	
Latin America	2.2	
Asia	1.9	
North America	0.7	
Europe	0.3	

Analysis

1. Which region has the fastest doubling time? Slowest doubling time?

2. How might this type of information be useful to governments of these regions?

3. What are some of the ecological implications for an area with a fast doubling time?

Name Date Class

Chapter 4

How can you determine the size of an animal population?

Preparation

Problem

How can you model a measuring technique to determine the size of an animal population?

Objectives

In this BioLab you will:

- **Model** the procedure used to measure an animal population.
- **Collect** data on a modeled animal population.
- **Calculate** the size of a modeled animal population.

Materials

paper bag containing beans
magic marker
calculator (optional)

Safety Precautions

Always wear goggles in the lab. Wash hands after working with plant material.

Skill Handbook

Use the **Skill Handbook** if you need additional help with this lab.

Procedure

1. Use the data table below.
2. Reach into your bag and remove 20 beans.
3. Use a dark magic marker to color these beans. These will represent your caught and marked animals.
4. When the ink has dried, return the beans to the bag.
5. Shake the bag. Without looking into the bag, reach in and remove 30 beans.
6. Record the number of marked beans (recaught and marked) and the number of unmarked beans (caught and unmarked) in your data table as trial 1.
7. Return all the beans to the bag.
8. Repeat steps 5 to 7 four more times for trials 2 to 5.
9. Calculate averages for each of the columns.
10. Using average values, calculate the original size of the bean population in the bag by using the following formula:

    M = number initially marked

    CwM = average number caught during the trials with marks

    Cw/oM = average number caught during the trials without marks

$$\text{Calculated Population Size} = \frac{M \times (CwM + Cw/oM)}{CwM}$$

11. Record the calculated population size in the data table.
12. To verify the actual population size, count the total number of beans in the bag and record this value in the data table.

How can you determine the size of an animal population?, *continued*

Chapter 4

Data Table

Trial	Total caught	Number caught with marks	Number caught without marks
1	20		
2	20		
3	20		
4	20		
5	20		
Totals	100		
Averages	20		

Calculated population size = ________

Actual population size = ________

Analyze and Conclude

1. **Thinking Critically** This experiment is a simulation. Explain why this type of activity is best done as a simulation.

2. **Applying Concepts** Give an example of how this technique could actually be used by a scientist.

3. **Analyzing Data** Compare the calculated to the actual population size. Explain why they may not agree exactly. What changes to the procedure would improve the accuracy of the activity?

4. **Making Inferences** Explain why this technique is used more often with animals than with plants when calculating population size.

5. **Making Predictions** Assume you were doing this experiment with living animals. What would you be doing in step 2? Step 3? Step 5?

MiniLab 5-1

Using Numbers

Measuring Species Diversity

Index of diversity (I.D.) is a mathematical way of expressing the amount of biodiversity and species distribution in a community. Communities with many different species (a high index of diversity) will be able to withstand environmental changes better than communities with only a few species (a low index of diversity.)

Procedure

1. Use the data table below.
2. Walk a city block or an area designated by your teacher and record the number of *different species* of trees present (you don't have to know their names, just that they differ by species). Record this number in your data table.
3. Walk the block or area again. This time, make a list of the trees by assigning each a number as you walk by it. Place an X under Tree 1 on your list. If Tree 2 is the same species as Tree 1, mark an X below it. Continue to mark an X under the trees as long as the species is the same as the previous one. When a different species is observed, mark an O under that tree on your list. Continue to mark an O if the next tree is the same species as the previous. If the next tree is different, mark an X.
4. Record in your data table:
   - **a.** the number of "runs." Runs are represented by a group of similar symbols in a row. Example—XXXXOOXO would be four runs (XXXX = 1 run, OO = 2 runs, X = 3, O = 4).
   - **b.** the total number of trees counted.
5. Calculate the Index of Diversity (I.D.) using the formula in the data table.

Data Table
Number of species =
Number of runs =
Number of trees =
Index of diversity = $\frac{\text{Number of species} \times \text{number of runs}}{\text{Number of trees}}$ =

Name Date Class

MiniLab 5-1

Measuring Species Diversity, *continued*

Analysis

1. Compare how your tree I.D. might compare with that of a vacant lot and with that of a grass lawn. Explain your answer.

2. If humans were concerned about biological diversity, would it be best to have a low or high I.D. for a particular environment? Explain your answer.

Name Date Class

MiniLab 5-2

Experimenting

Conservation of Soil

Soil is an important natural resource and should be conserved. How does one conserve soil? As a start, we should be aware of factors that promote or speed up its unnecessary loss or erosion.

Procedure

1 Use the data table below.

Data Table

Source of sample	Volume of original water	Volume of collected water	Volume of eroded soil
Bare soil			
Soil with grass			

2 Measure 200 mL of water in a beaker.

3 Fill a plastic tray with soil as shown on page 126 of your text.

4 Pour the water onto the soil, tilting the tray and placing a dish underneath the end of it as indicated in the diagram on page 126.

5 Wait several minutes for all water and soil to drain into the dish.

6 Pour the soil and water from the dish into a graduated cylinder. Wait several minutes for the soil and water to settle. Measure the volume of soil and water that washed or eroded into the dish. Record these values in your data table.

7 Repeat steps 2–6, only this time place a section of soil in which grass is growing onto the tray. **CAUTION:** ***Always wash your hands with soap and water after working with soil.***

Analysis

1. What part of the experiment simulated soil erosion?

2. Based on this experiment, explain why farmers usually plant unused fields with some type of crop cover.

Name Date Class

Researching Information on Exotic Pets

Chapter 5

Preparation

Problem

How can you use the Internet to gather information on keeping an exotic animal as a pet?

Objectives

In this BioLab, you will:

- **Select** on animal that is considered an exotic pet.
- **Use the Internet** to collect and compare information from other students.
- **Conclude** if the animal you have chosen would or would not make a good pet.

Materials

access to the Internet

Procedure

1. Use the data table as a guide for the information to be researched.
2. Pick an exotic pet from the following list of choices: hedgehog, snake, ferret, large cat such as a tiger or panther, monkey, ape, iguana, tropical bird.
3. Go to the Glencoe Science Web Site at **www.glencoe.com/sec/science to find links** that will provide you with information for this BioLab. Note: You are not limited to the pet suggestions provided. If a different organism appeals to you, research it instead.
4. Record your findings in the data table.

Data Table

Category	Response
Exotic pet choice	
Scientific name	
Natural habitat (where found in nature)	
Adult size	
Dietary needs	
Special health problems	
Source of medical care, if needed	
Safety issues for humans	
Size of cage area needed	
Special environmental needs	
Social needs	
Cost of purchase	
Cost of maintaining (monthly estimate)	
Care issues (high/low maintenance)	
Additional information	
Additional sources	

Name Date Class

Chapter 5

Researching Information on Exotic Pets, *continued*

ANALYZE AND CONCLUDE

Analyze and Conclude

1. **Defining Operationally** What is meant by the term *domesticated*?

2. **Using the Internet** Look at the findings posted by other students. Which of the animals researched would make the best pet? Which would not be a wise choice? Explain.

3. **Interpreting Data** What do you consider the most important information gained from your research that:
   a. supports keeping your exotic pet choice?

   b. does not support keeping your exotic pet choice?

4. **Thinking Critically** What positive contribution might be made to the cause of conservation when keeping an exotic pet? Explain.

5. **Thinking Critically** How can keeping exotic pets be a negative influence on conservation biology efforts?

6. **Analyzing Information** What are some reasons why zoos rather than individuals are better able to handle exotic animals?

Name Date Class

MiniLab 6-1 Determine pH

Experimenting

The pH of a solution is a measurement of how acidic or basic that solution is. An easy way to measure the pH of a solution is to use pH paper.

Procedure

1. Pour a small amount (about 5 mL) of each of the following into separate clean, small beakers or other small glass containers: lemon juice, household ammonia solution, liquid detergent, shampoo, and vinegar.
2. Dip a fresh strip of pH paper briefly into each solution and remove.
3. Compare the color of the wet paper with the pH color chart; record the pH of each material. **CAUTION:** ***Wash your hands after handling lab materials.***

Analysis

1. Which solutions are acids?

___

___

2. Which solutions are bases?

___

___

3. What ions in the solution caused the pH paper to change? Which solution contained the highest concentration of hydroxide ions? How do you know?

___

___

MiniLab 6-2

Examine the Rate of Diffusion

Applying Concepts

In this lab, you will place a small potato cube in a solution of potassium permanganate and observe how far the dark purple color diffuses into the potato after a given length of time.

Procedure

1. Using a single-edge razor blade, cut a cube 1 cm on each side from a raw, peeled potato. **CAUTION:** ***Be careful with sharp objects.*** **Do not cut objects while holding them in your hand.**
2. Place the cube in a cup or beaker containing the purple solution. The solution should cover the cube. Note and record the time. Let the cube stand in the solution for between 10 and 30 minutes.
3. Using forceps, remove the cube from the solution and note the time. Cut the cube in half.
4. Measure, in millimeters, how far the purple solution has diffused, and divide this number by the time you allowed your potato to remain in the solution. This is the diffusion rate.

Analysis

1. How far did the purple solution diffuse?

____

____

2. What was the rate of diffusion per minute?

____

____

Name Date Class

Chapter 6

Does temperature affect an enzyme reaction?

PREPARATION

Problem

Does the enzyme peroxidase work in cold temperatures? Does peroxidase work better at higher temperatures? Does peroxidase work after being frozen or boiled?

Hypotheses

Make a hypothesis regarding how you think temperature will affect the rate at which the enzyme peroxidase breaks down hydrogen peroxide. Consider both low and high temperatures.

Objectives

In this BioLab, you will:

- **Observe** the activity of an enzyme.
- **Compare** the activity of the enzyme at various temperatures.

Possible Materials

clock or timer
400-mL beaker
kitchen knife
tongs or large forceps
5-mm thick potato slices
3% hydrogen peroxide
ice
hot plate
waxed paper
thermometer

Safety Precautions

Be sure to wash your hands before and after handling the lab materials. Always wear goggles in the lab.

Skill Handbook

Use the **Skill Handbook** if you need additional help with this lab.

PLAN THE EXPERIMENT

1. Decide on a way to test your group's hypothesis. Keep the available materials in mind.
2. When testing the activity of the enzyme at a certain temperature, consider the length of time it will take for the potato to reach that temperature, and how the temperature will be measured.
3. To test for peroxidase activity, add 1 drop of hydrogen peroxide to the potato slice and observe what happens.
4. When heating a thin potato slice, first place it in a small amount of water in a beaker. Then heat the beaker slowly so that the temperature of the water and the temperature of the slice are always the same. Try to make observations at several temperatures between 10°C and 100°C.

Check the Plan

Discuss the following points with other groups to decide on the final procedure for your experiment.

1. What data will you collect? How will you record them?
2. What factors should be controlled?
3. What temperatures will you test?
4. How will you achieve those temperatures?
5. ***Make sure your teacher has approved your experimental plan before you proceed further.***
6. Carry out your experiment. **CAUTION:** ***Be careful with chemicals and heat. Wash your hands after the lab.***

Name Date Class

Chapter 6

Does temperature affect an enzyme reaction?, *continued*

ANALYZE AND CONCLUDE

Analyze and Conclude

1. **Checking Your Hypothesis** Do your data support or reject your hypothesis? Explain your results.

2. **Analyzing Data** At what temperature did peroxidase work best?

3. **Identifying Variables** What factors did you need to control in your tests?

4. **Recognizing Cause and Effect** If you've ever used hydrogen peroxide as an antiseptic to treat a cut or scrape, you know that it foams as soon as it touches an open wound. How can you account for this observation?

Name Date Class

MiniLab 7-1

Measuring in SI

Measuring Objects Under a Microscope

Knowing the diameter of the circle of light you see when looking through a microscope allows you to measure the size of objects that are being viewed. For most microscopes, the diameter of the circle of light is 1.5 mm, or 1500 μm (micrometers), under low power and 0.375 mm, or 375 μm, under high power.

Refer to *Practicing Scientific Methods* in the **Skill Handbook** if you need help with SI units.

Procedure

1 Look at diagram A on page 177 of your text that shows an object viewed under low power. Knowing the circle diameter to be 1500 μm, the estimated length of object (a) is 400 μm. What is the estimated length of object (b)?

___

2 Look at diagram B on page 177 of your text that shows an object viewed under high power. Knowing the circle diameter to be 375 μm, the estimated length of object (c) is 100 μm. What is the estimated length of object (d)?

___

3 Prepare a wet mount of a strand of your hair. Your teacher can help with this procedure. **CAUTION:** ***Use caution when handling microscopes and glass slides.*** Measure the width of your hair strand while viewing it under low and then high power.

___

Analysis

1. An object can be magnified 100, 200, or 1000 times when viewed under a microscope. Does the object's actual size change with each magnification? Explain.

___

___

___

2. Do your observations of the size of your hair strand under low and high power support the answer to question 1? If not, offer a possible explanation why.

___

___

___

Name Date Class

MiniLab 7-2

Experimenting

Cell Organelles

Adding stains to cellular material helps you distinguish cell organelles.

Procedure

CAUTION: ***Be sure to wash hands before and after this experiment.***

1 Prepare a water wet mount of onion skin. Do this by using your fingernail to peel off the inside of a layer of onion bulb. The layer must be almost transparent. Use the diagram on page 187 of your text as a guide.

2 Make sure that the onion layer is lying flat on the glass slide and not folded.

3 Observe the onion cells under low- and high-power magnification. Identify as many organelles as possible.

4 Repeat steps 1 through 3, only this time use an iodine stain instead of water.

Analysis

1. What organelles were easily seen in the unstained onion cells? Cells stained with iodine?

---

2. How are stains useful for viewing cells?

Name Date Class

Observing and Comparing Different Cell Types

Chapter 7

Preparation

Problem

Are all cells alike in appearance and size?

Objectives

In this BioLab, you will:

- **Observe**, **diagram**, and **measure** cells and their organelles.
- **Hypothesize** which cells are from prokaryotes, eukaryotes, unicellular organisms, and multicellular organisms.
- **List** the traits of plant and animal cells.

Materials

microscope
glass slide
water
dropper
coverslip
forceps
prepared slides of *Bacillus subtilus*, frog blood, and *Elodea*

Safety Precautions

Always wear goggles in the lab.

Skill Handbook

Use the **Skill Handbook** if you need additional help with this lab.

Procedure

1. Use the data table below.
2. Examine a prepared slide of *Bacillus subtilus* using both low- and high-power magnification. (NOTE: This slide has been stained. Its natural color is clear.)
   CAUTION: ***Use care when handling slides. Dispose of any broken glass in a container provided by your teacher.***

Data Table

	Bacillus subtilus	*Elodea*	Frog Blood
Organelles observed			
Prokaryote or eukaryote			
From a multicellular or unicellular organism			
Diagram (with size in micrometers, µm)			

Name Date Class

Observing and Comparing Different Cell Types, *continued*

Chapter 7

3. Look for and record the names of any observed organelles. Hypothesize if these cells are prokaryotes or eukaryotes. Hypothesize if these cells are from a unicellular or multicellular organism. Record your findings in the table.
4. Diagram one cell as seen under high-power magnification.
5. While using high power, determine the length and width in micrometers of this cell. Refer to *Practicing Scientific Methods* in the **Skill Handbook** for help with determining magnification. Record your measurements on the diagram.
6. Prepare a wet mount of a single leaf from *Elodea* using the diagram as a guide.
7. Observe cells under low- and high-power magnification.
8. Repeat steps 3 through 5 for *Elodea*.
9. Examine a prepared slide of frog blood. (NOTE: This slide has been stained. Its natural color is pink.)
10. Observe cells under low- and high-power magnification.
11. Repeat steps 3 through 5 for frog blood cells.

Analyze and Conclude

1. **Observing and Inferring** Which cells were prokaryotes? How were you able to tell?

2. **Observing and Inferring** Which cells were eukaryotes? How were you able to tell?

3. **Predicting** Which cell was from a plant, from an animal? Explain your answer.

4. **Measuring** Are prokaryote or eukaryote cells larger? Give specific measurements to support your answer.

5. **Defining Operationally** Describe how plant and animal cells are alike and how they differ.

Name Date Class

MiniLab 8-1

Formulating Models

Cell Membrane Simulation

If membranes show selective permeability, what might happen if a plastic bag (representing a cell's membrane) were filled with starch molecules on the inside and surrounded by iodine molecules on the outside?

Procedure

1. Fill a plastic bag with 50 mL of starch. Seal the bag with a twist tie.
2. Fill a beaker with 50 mL of of iodine solution. **CAUTION:** ***Rinse with water if iodine gets on skin. Iodine is toxic.***
3. Note and record the color of the starch and iodine.
4. Place the bag into the beaker. **CAUTION:** ***Wash your hands after handling lab materials.***
5. Note and record the color of the starch and iodine 24 hours later.

Analysis

1. Describe and compare the color of the iodine and starch at the start and at the conclusion of the experiment.

____

2. Fact: Starch mixed with iodine forms a purple color.

a. In which direction did the iodine move? What is your evidence?

____

b. In which direction did the starch move? What is your evidence?

____

____

____

3. Explain how this experiment illustrates selective permeability.

____

____

____

Name Date Class

MiniLab 8-2

Comparing and Contrasting

Seeing Asters

The result of the process of mitosis is similar in plant and animal cells. However, animal cells have asters whereas plant cells do not. Animal cells undergoing mitosis clearly show these structures.

Procedure

1. Examine a slide marked "fish mitosis" under low- and high-power magnification. **CAUTION:** ***Use care when handling prepared slides.***
2. Find cells that are undergoing mitosis. You will be able to see dark-stained rodlike structures within certain cells. These structures are chromosomes.
3. Note the appearance and location of asters. They will appear as ray or starlike structures at opposite ends of cells that are in metaphase.
4. Asters may also be observed in cells that are in other phases of mitosis.

Analysis

1. Describe the appearance and location of asters in cells that are in prophase.

____

____

2. Explain how you know that asters are not critical to mitosis.

____

____

3. Design an experiment that tests the hypothesis that asters are not essential for mitosis in animal cells.

____

____

____

____

____

Name Date Class

Chapter 8

Where is mitosis most common?

PREPARATION

Problem

Does mitosis occur at the same rate in all parts of an onion root?

Objectives

In this BioLab, you will:

- **Observe** cells in two different root areas.
- **Identify** the stages of mitosis in each area.

Materials

prepared slide of onion root tip
microscope

Skill Handbook

Use the **Skill Handbook** if you need additional help with this lab.

PROCEDURE

1. Use the data table below.
2. Using Diagram **A** on page 221 of your text as a guide, locate area X on a prepared slide of onion root tip.
3. Place the prepared slide under your microscope and use low power to locate area X. **CAUTION:** ***Use care when handling prepared slides.***
4. Switch to high power.
5. Using Diagram **B** on page 221 of your text as a guide:
   a. Identify those cells that are in mitosis and in interphase.
   b. Record in the data table the number of cells observed in each phase of mitosis and interphase for area X.

   (NOTE: It will be easier to count and keep track of cells by following rows. See Diagram C on page 221 of your text as a guide to counting.)
6. Using Diagram **A** again, locate area Y on the same prepared slide.
7. Place the prepared slide under your microscope and use low power to locate area Y.
8. Switch to high power.
9. Using Diagram **B** as a guide:
   a. Identify those cells that are in mitosis and in interphase.
   b. Record in the data table the number of cells observed in each phase of mitosis and interphase for area Y.

Data Table

Phase	Area X	Area Y
Interphase		
Prophase		
Metaphase		
Anaphase		
Telophase		

Name Date Class

Chapter 8

Where is mitosis most common?, *continued*

Analyze and Conclude

1. **Observing** Which area of the onion root tip (X or Y) had the greatest number of cells undergoing mitosis? The fewest? Use specific totals from your data table to support your answer.

2. **Predicting** If mitosis is associated with rapid growth, where do you believe is the location of most rapid root growth, area X or Y? Explain your answer.

3. **Applying** Where might you look for cells in the human body that are undergoing mitosis?

4. **Calculating** According to your data, which phase of mitosis is most common? Least common?

5. **Critical Thinking** Assume that you were not able to observe cells in every phase of mitosis. Explain why this might be.

Name Date Class

MiniLab 9-1

Formulating Models

Use Isotopes to Understand Photosynthesis

C.B. van Niel discovered the steps of photosynthesis when he used radioactive isotopes of oxygen as tracers. Radioactive isotopes are used to follow a particular molecule through a chemical reaction.

Procedure

1. Study the following equation for photosynthesis that resulted from the van Niel experiment:

$$6CO_2 + 6H_2O^* \rightarrow C_6H_{12}O_6 + 6O_2^*$$

2. Radioactive water, water tagged with an isotope of oxygen as a tracer (shown with the *), was used. Note where the tagged oxygen in water ends up on the right side of the chemical reaction.
3. Assume that van Niel repeated his experiment, but this time he put a radioactive tag on the oxygen in CO_2.
4. Using materials provided by your teacher, model what you would predict the appearance of his results would be. Your model must include a "tag" to indicate the oxygen isotope on the left side of the arrow as well as where it ends up on the right side of the arrow.
5. You must use labels or different colors in your model to indicate also the fate of carbon and hydrogen.

Analysis

1. Explain how an isotope can be used as a tag.

____

____

2. Using your model, predict:

a. the fate of all oxygen molecules that originated from carbon dioxide.

____

____

b. the fate of all carbon molecules that originated from carbon dioxide.

____

____

c. the fate of all hydrogen molecules that originated from water.

____

____

Name Date Class

MiniLab 9-2

Predicting

Determine if Apple Juice Ferments

Organisms such as yeast have the ability to break down food molecules and synthesize ATP when no oxygen is available. When the appropriate food is available, yeast can carry out alcoholic fermentation, producing CO_2. Thus, the production of CO_2 can be used to judge whether alcoholic fermentation is taking place.

Procedure

1 Carefully study the diagram on page 242 of your text and set up the experiment as shown.

2 Hold the test tube in a beaker of warm (not hot) water and observe.

Analysis

1. What were the gas bubbles that came from the plastic pipette?

2. Predict what would happen to the rate of bubbles given off if more yeast were present in the mixture.

3. Why was the test tube placed in warm water?

4. On the basis of your observations, was this process aerobic or anaerobic?

Name Date Class

Chapter 9

What factors influence photosynthesis?

Preparation

Problem

How do different wavelengths of light a plant receives affect its rate of photosynthesis?

Objectives

In this BioLab, you will:

- **Observe** photosynthesis in an aquatic organism.
- **Measure** the rate of photosynthesis.
- **Observe** how various wavelengths of light influence the rate of photosynthesis.
- **Use the Internet** to collect and compare data from other students.

Materials

1000-mL beaker
three *Elodea* plants
string
washers
colored cellophane, assorted colors
lamp with reflector and 150-watt bulb
0.25 percent sodium hydrogen carbonate (baking soda) solution
watch with second hand

Safety Precautions

Always wear goggles in the lab.

Skill Handbook

Use the **Skill Handbook** if you need additional help with this lab.

Procedure

1. Construct a basic setup like the one shown on page 244 of your text.
2. Use the data table below to record your measurements. Be sure to include a column for each color of light you will investigate and a column for the control experiment.
3. Place the *Elodea* plants in the beaker, then completely cover the plants with water. Add some of the baking soda solution. The solution provides CO_2 for the aquarium plants. **Be sure to use the same amount of water and solution for each trial.**
4. Conduct a control experiment by directing the lamp (without colored cellophane) on the plant and notice when you see the bubbles.
5. Observe and record the number of oxygen bubbles that *Elodea* generates in five minutes.
6. Repeat steps 4 and 5 with a piece of colored cellophane. Record your observations.
7. Repeat steps 4 and 5 with a different color of cellophane and record your observations.
8. Go to the Glencoe Science Web Site at **www.glencoe.com/sec/science** to **post your data.**

Data Table

	Control	Color 1	Color 2
Bubbles observed in five minutes			

Name Date Class

Chapter 9

What factors influence photosynthesis?, *continued*

Analyze and Conclude

1. **Interpreting Observations** From where did the bubbles of oxygen emerge? Why?

2. **Making Inferences** Explain how counting bubbles measures the rate of photosynthesis.

3. **Using the Internet** Make a graph of your data and data posted by other students with the rate of photosynthesis per minute plotted against the wavelength of light you tested for both the control and experimental setups. Write a sentence or two explaining the graph.

MiniLab 10-1

Observing and Inferring

Looking at Pollen

Pollen grains are formed within the male anthers of flowers. What is their role? Pollen contains the male gametes or sperm cells needed for fertilization. This means that pollen grains carry the hereditary units from male parent plants to female parent plants. The pollen grains that Mendel transferred from the anther of one pea plant to the pistil of another plant carried the hereditary traits that he so carefully observed in the next generation.

Procedure

1. Examine a flower. Using the diagram on page 260 of your text as a guide, locate the stamens of your flower. There are usually several stamens in each flower.
2. Remove one stamen and locate the enlarged end—the anther.
3. Add a drop of water to a microscope glass slide. Place the anther in the water. Add a coverslip. Using the eraser end of a pencil, tap the coverslip several times to squash the anther.
4. Observe under low power. Look for numerous small round structures. These are pollen grains.

Analysis

1. Provide an estimate of the number of pollen grains present in an anther.

2. Describe the appearance of a single pollen grain.

3. Explain the role of pollen grains in plant reproduction.

Name Date Class

MiniLab 10-2

Formulating Models

Modeling Crossing Over

Crossing over occurs during meiosis and involves only the nonsister chromatids that are present during tetrad formation. The process is responsible for the appearance of new combinations of alleles in gamete cells.

Procedure

1. Use the data table below.
2. Roll out four long strands of clay at least 10 cm long to represent two chromosomes, each with two chromatids.
3. Use the figure on page 274 of your text as a guide in joining and labeling these model chromatids. Although there are four chromatids, assume that they started out as a single pair of homologous chromosomes prior to replication. The figure shows tetrad formation during prophase I of meiosis.
4. First, assume that no crossing over takes place. Model the appearance of the four gamete cells that will result at the end of meiosis. Record your model's appearance by drawing the gametes' chromosomes and their genes in your data table.
5. Next, repeat steps 2–4. This time, however, assume that crossing over occurs between genes B and C.

Data Table

No crossing over	Crossing Over
Appearance of gamete cells	Appearance of gamete cells

Name Date Class

MiniLab 10-2

Modeling Crossing Over, *continued*

Analysis

1. Predict and diagram the appearance of the chromosomes prior to replication.

2. Define crossing over and explain when it occurs.

3. Compare any differences in the appearance of genes on chromosomes in gamete cells when crossing over occurs and when it does not occur.

4. Crossing over has been compared to “shuffling the deck” in cards. Explain what this means.

5. What would be accomplished if crossing over occurred between sister chromatids? Explain your answer.

Name Date Class

Chapter 10

How can phenotypes and genotypes of plants be determined?

PREPARATION

Problem

Can the phenotypes and genotypes of the parent plants that produced two groups of seeds be determined from the phenotypes of the plants grown from the seeds?

Hypotheses

Have your group agree on a hypothesis to be tested that will answer the problem question. Record your hypothesis.

Objective

In this BioLab, you will:

- **Analyze** the results of growing two groups of seeds.
- **Draw conclusions** about phenotypes and genotypes based on those results.
- **Use the Internet** to compare data from other schools.

Possible Materials

potting soil
small flowerpots or seedling flats
two groups of tobacco seeds
hand lens
light source
thermometer
plant-watering bottle

Safety Precautions

CAUTION: ***Always wash your hands after handling plant materials.***

Skill Handbook

Use the **Skill Handbook** if you need additional help with this lab.

PLAN THE EXPERIMENT

1. Examine the materials provided by your teacher. As a group, make a list of the possible ways you might test your hypothesis.
2. Agree on one way that your group could investigate your hypothesis.
3. Design an experiment that will allow you to collect quantitative data. For example, how many plants do you think you will need to examine?
4. Prepare a numbered list of directions. Include a list of materials and the quantities you will need.
5. Make a data table for recording your observations.

Check the Plan.

1. Carefully determine what data you are going to collect. How many seeds do you think you will need? How long will you carry out the experiment?
2. What variables, if any, will have to be controlled? (Hint: Think about the growing conditions for the plants.)
3. ***Make sure your teacher has approved your experimental plan before you proceed.***
4. Carry out your experiment. Make any needed observations, such as the numbers of green and albino plants in each group, and complete your data table.
5. **Post your data** on the Glencoe Science Web Site at:

   www.glencoe.com/sec/science

Name Date Class

How can phenotypes and genotypes of plants be determined?, *continued*

Chapter 10

Analyze and Conclude

1. **Thinking Critically** Why was is necessary to grow plants from the seeds in order to determine the phenotypes of the plants that formed the seeds?

2. **Drawing Conclusions** Using the information in the introduction on page 280 of your text, describe how the gene for green color (*C*) is inherited.

3. **Making Inferences** For the group of seeds that yielded all green plants, are you able to determine exactly the genotypes of the parents that formed these seeds? Can you determine the genotype of each plant observed? Explain.

4. **Making Inferences** For the group of seeds that yielded some green and some albino plants, are you able to determine exactly the genotypes of the plants that formed these seeds? Can you determine the genotype of each plant observed? Explain.

5. **Using the Internet** Compare your experimental design with that of other students. Were your results similar? What might account for the differences?

Name Date Class

MiniLab 11-1

Predicting

Transcribe and Translate

Molecules of DNA carry the genetic instructions for protein formation. Converting these DNA instructions into proteins requires a series of coordinated steps in transcription and translation.

Procedure

1. Use the data table below.
2. Complete column B by writing the correct mRNA codon for each sequence of DNA bases listed in the column marked *DNA Base Sequence*. Use the letters A, U, C, or G.
3. Identify the process responsible by writing its name on the arrow in column A.
4. Complete column D by writing the correct anticodon that bonds to each codon from column B.
5. Identify the process responsible by writing its name on the arrow in column C.
6. Complete column E by writing the name of the correct amino acid that is coded by each base sequence. Use *Table 11.2* on page 298 of your text to translate the mRNA base sequences to amino acids.

Data Table

	A	B	C	D	E
DNA base sequence	Process	mRNA codon	Process	tRNA anticodon	Amino acid
AAT	——→		——→		
GGG	——→		——→		
ATA	——→		——→		
AAA	——→		——→		
GTT	——→		——→		

MiniLab 11-1

Transcribe and Translate, *continued*

Analysis

1. Where within the cell:

a. are the DNA instructions located?

b. does transcription occur?

c. does translation occur?

2. Describe the structure of a tRNA molecule.

3. Explain why specific base pairing is essential to the processes of transcription and translation.

Name Date Class

MiniLab 11-2

Making and Using Tables

Gene Mutations and Proteins

Gene mutations often have serious effects on proteins. In this activity, you will demonstrate how such mutations affect protein synthesis.

Procedure

1. Use the following base sequence of one strand of an imaginary DNA molecule: AATGCCAGTGGTTCGCAC.
2. Write the base sequence for an mRNA strand that would be transcribed from the given DNA sequence.
3. Use *Table 11.2* to determine the sequence of amino acids in the resulting protein fragment.
4. If the fourth base in the original DNA strand were changed from G to C, how would this affect the resulting protein fragment?
5. If a G were added to the original DNA strand after the third base, what would the resulting mRNA look like? How would this addition affect the protein?

Analysis

1. Which change in DNA was a point mutation? Which was a frameshift mutation?

____

____

2. In what way did the point mutation affect the protein?

____

____

3. How did the frameshift mutation affect the protein?

____

____

Name Date Class

Chapter 11

RNA Transcription

PREPARATION

Problem

How does the order of bases in DNA determine the order of bases in mRNA?

Objectives

In this BioLab, you will:

- **Formulate a model** to show how the order of bases in DNA determines the order of bases in mRNA.
- **Infer** why the structure of DNA enables it to be easily copied.

Materials

construction paper, 5 colors
scissors
clear tape
pencil

Safety Precautions

Be careful when using scissors. Always wear goggles in the lab.

Skill Handbook

Use the **Skill Handbook** if you need additional help with this lab.

PROCEDURE

1. Copy the illustrations of the four different DNA nucleotides on page 308 of your text onto your construction paper, making sure that each different nucleotide is on a different color paper. You should make ten copies of each nucleotide.
2. Using scissors, carefully cut out the shapes of each nucleotide.
3. Using any order of nucleotides that you wish, construct a double-stranded DNA molecule. If you need more nucleotides, copy them as before.
4. Fasten your molecule together using clear tape. Do not tape across base pairs.
5. As in step 1, copy the illustrations of A, G, and C nucleotides. Use the same colors of construction paper as in step 1. Use the fifth color of construction paper to make copies of uracil nucleotides.
6. With scissors, carefully cut out the nucleotide shapes.
7. With your DNA molecule in front to you, demonstrate the process of transcription by first pulling the DNA molecule apart between the base pairs.
8. Using only one of the strands of DNA, begin matching complementary mRNA nucleotides with the exposed bases on the DNA model to make RNA.
9. When you are finished, tape your new mRNA molecule together.

Name Date Class

Chapter 11

RNA Transcription, *continued*

Analyze and Conclude

1. **Observing and Inferring** Does the mRNA model more closely resemble the DNA strand from which it was transcribed or the complementary strand that wasn't used? Explain your answer.

2. **Recognizing Cause and Effect** Explain how the structure of DNA enables the molecule to be easily transcribed. Why is this important for genetic information?

3. **Relating Concepts** Why is RNA important to the cell? How does an mRNA molecule carry information from DNA?

DNA Nucleotides for BioLab, Chapter 11 (reproduce 4 copies per student)

Copy these onto white paper and have students color the models with colored pencils—a different color for each nucleotide.

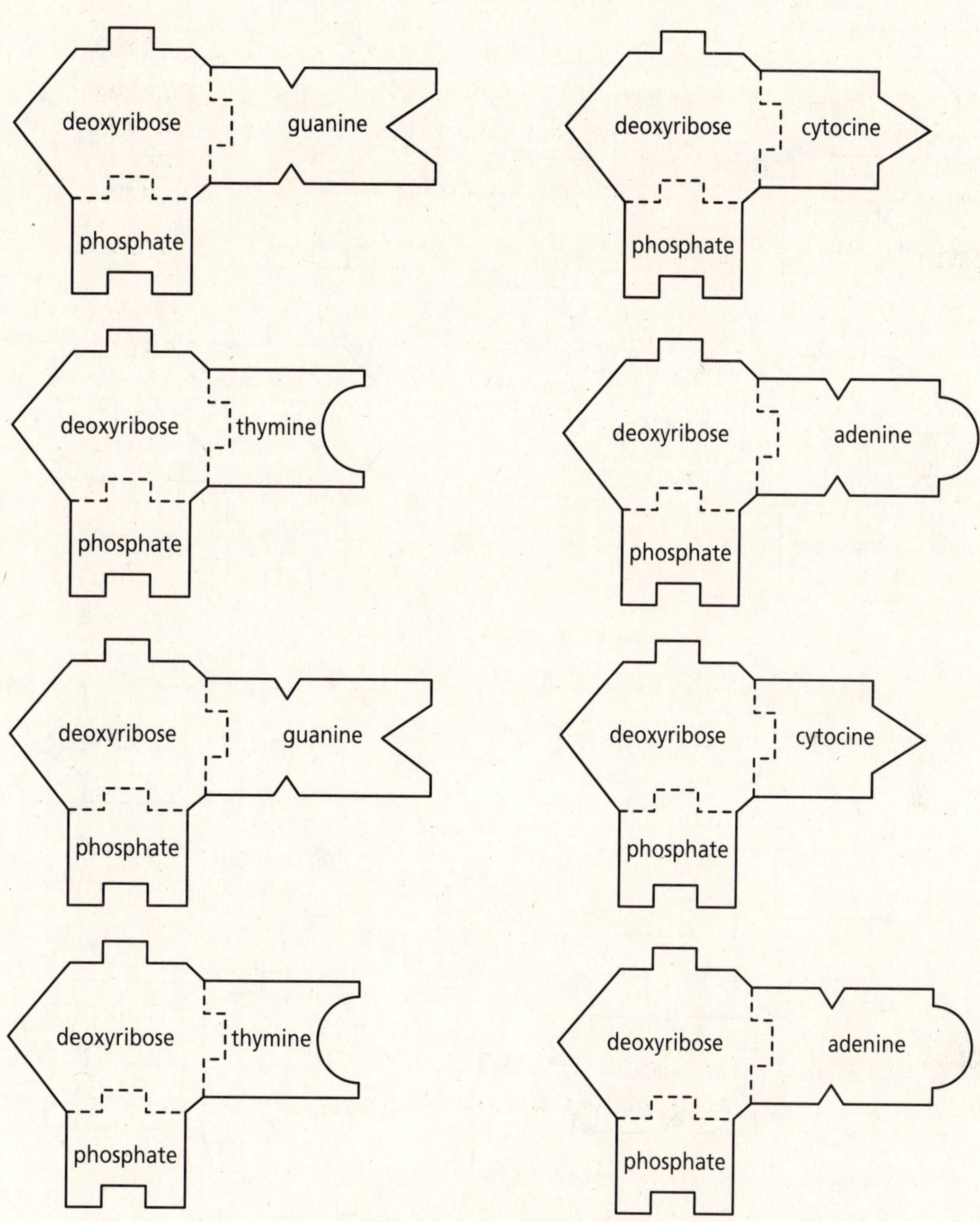

RNA Nucleotides for BioLab, Chapter 11
(reproduce 2 copies per student)

Copy these onto white paper and have students color the models with colored pencils—a different color for each nucleotide.

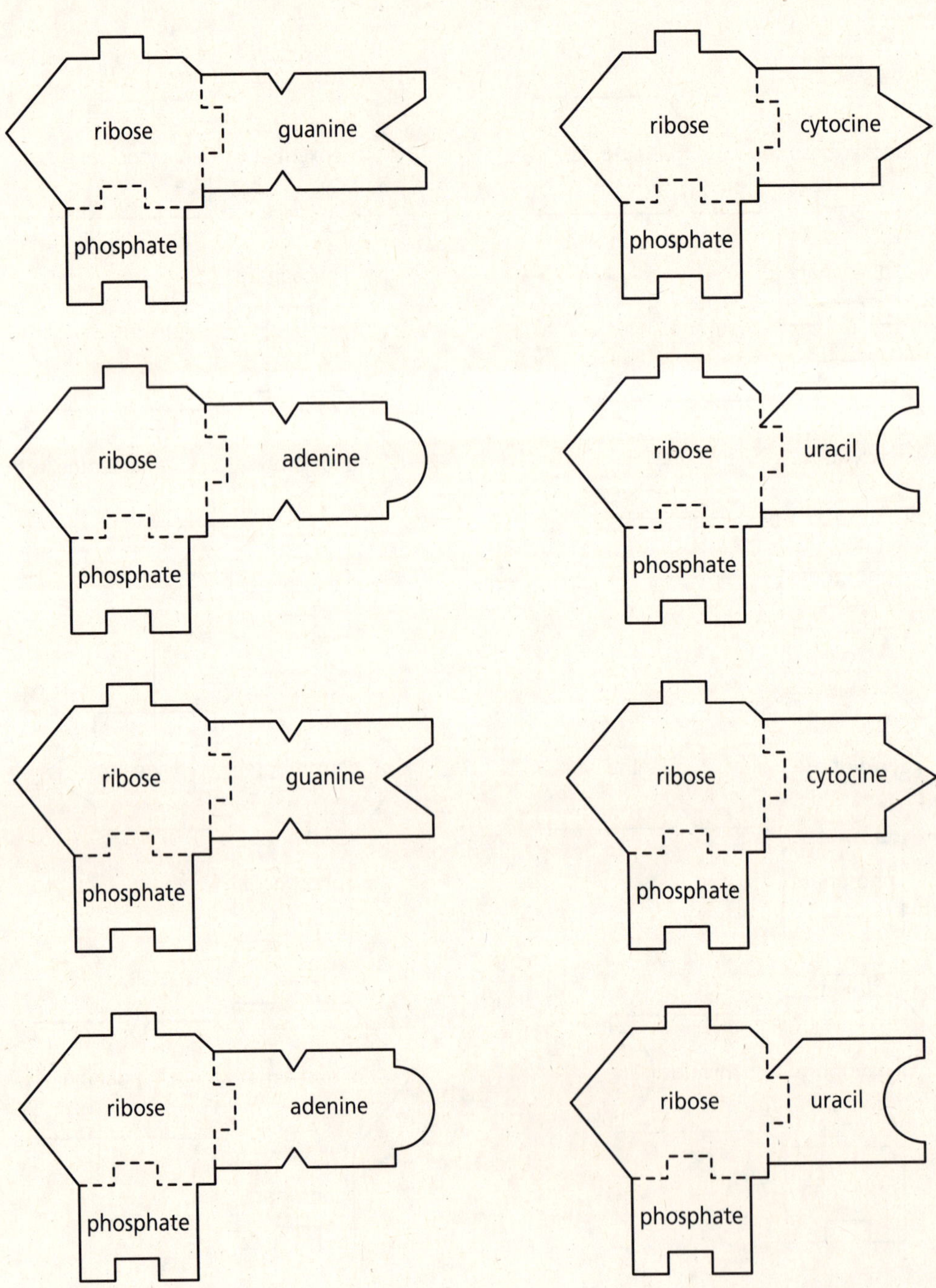

Name Date Class

MiniLab 12-1

Analyzing Information

Illustrating a Pedigree

The pedigree method of studying a trait in a family uses records of phenotypes extending for two or more generations. Studies of pedigrees can be used to yield a great deal of genetic information about a related group.

Procedure

1. Working with a partner, choose one human trait, such as attached and free-hanging ear-lobes or tongue rolling, that interests both of you.
2. Using either your or your partner's family, collect information about your chosen trait. Include whether each individual is male or female, does or does not have the trait, and the relationship of the individual to others in the family.
3. Use your information to draw a pedigree for the trait.
4. Try to determine how your trait is inherited.

Analysis

1. What trait did you study? Can you determine from your pedigree what the apparent inheritance pattern of the trait is?

____

____

____

2. How is the study of inheritance patterns limited by pedigree analysis?

____

____

____

MiniLab 12-2

Observing and Inferring

Detecting Colors and Patterns in Eyes

Human eye color, like skin color, is determined by polygenic inheritance. You can detect several shades of eye color, especially if you look closely at the iris with a magnifying glass. Often, the pigment is deposited so that light reflects from the eye, causing the iris to appear blue, green, gray, or hazel (brown-green). In actuality, the pigment may be yellowish or brown, but not blue.

Procedure

1. Use a magnifying glass to observe the patterns and colors of pigment in the eyes of five classmates.
2. Use colored pencils to make drawings of the five irises.
3. Describe your observations in your journal.

Analysis

1. How many different pigments were you able to detect in each eye?

2. From your data, do you suspect that eye color might not be inherited by simple Mendelian rules? Explain.

3. Suppose that two people have brown eyes. They have two children with brown eyes, one with blue eyes, and one with green eyes. What pattern might this suggest?

Name Date Class

Chapter 12

What is the pattern of cytoplasmic inheritance?

Preparation

Problem

What inheritance pattern does the variegated leaf trait in *Brassica* show?

Hypotheses

Consider the possible evidence you could collect that would answer the problem question. Among the people in your group, form a hypothesis that you can test to answer the question, and write the hypothesis in your journal.

Objectives

In this BioLab, you will:

- **Determine** which crosses of *Brassica* plants will reveal the pattern of cytoplasmic inheritance.
- **Analyze** data from *Brassica* crosses.

Possible Materials

Brassica rapa seeds, normal and variegated
potting soil and trays
paintbrushes
forceps
single-edge razor blade
light source
labels

Safety Precautions

Always wear goggles in the lab. Handle the razor blade with extreme caution. Always cut away from you. Wash your hands with soap and water after working with plant material.

Skill Handbook

Use the **Skill Handbook** if you need additional help with this lab.

Plan the Experiment

1. Decide which crosses will be needed to test your hypothesis.
2. Keep the available materials in mind as you plan your procedure. How many seeds will you need?
3. Record your procedure, and list the materials and quantities you will need.
4. Assign a task to each member of the group. One person should write data in a journal, another can pollinate the flowers, while a third can set up the plant trays. Determine who will set up and clean up materials.
5. Design and construct a data table for recording your observations.

Check the Plan

Discuss the following points with other group members to decide the final procedure for your experiment.

1. What data will you collect, and how will data be recorded?
2. When will you pollinate the flowers? How many flowers will you pollinate?
3. How will you transfer pollen from one flower to another?
4. How and when will you collect the seeds that result from your crosses?
5. What variables will have to be controlled? What controls will be used?
6. When will you end the experiment?
7. ***Make sure your teacher has approved your experimental plan before you proceed further.***
8. Carry out your experiment.

DESIGN YOUR OWN BioLab

What is the pattern of cytoplasmic inheritance?, *continued*

Chapter 12

Analyze and Conclude

1. **Checking Your Hypothesis** Did your data support your hypothesis? Why or why not?

2. **Interpreting Observations** What is the inheritance pattern of variegated leaves in *Brassica?*

3. **Making Inferences** Explain why genes in the chloroplast are inherited in this pattern.

4. **Drawing Conclusions** Which parent is responsible for passing the variegated trait to its offspring?

5. **Making Scientific Illustrations** Draw a diagram tracing the inheritance of this trait through cell division.

Name Date Class

MiniLab 13-1

Applying Concepts

Matching Restriction Enzymes to Cleavage Sites

Many restriction enzymes cut sequences of DNA that are palindromes. As a result of cuts to the DNA, single-stranded sequences of DNA are left dangling at the ends of a fragment. These ends are available for pairing with their complementary bases in a plasmid or piece of viral DNA.

Procedure

1 Use the data table below.

Data Table

DNA fragment	Enzyme letter (D–F)	Action of restriction enzyme	Cleaved fragment of DNA
—GGTACC— |||||| —CCATGG—	E	—G|GTACC— |- - - -,| —CCATG|G—	—G GTACC— | | —CCATG G—
—CCATGG— |||||| —GGTACC—			
—CAATTG— |||||| —GTTAAC—			
—GATATC— |||||| —CTATAG—			

2 Figure out which restriction enzyme will cleave each DNA fragment. Use the following guides.

Enzyme D cleaves at an A-A site and leaves 3 single-stranded bases on each end.

Enzyme E cleaves at a G-G site and leaves 4 single-stranded bases on each end.

Enzyme F cleaves at a G-A site and leaves 4 single-stranded bases on each end.

3 Draw in the action of each enzyme. Record its letter.

4 Diagram each fragment of DNA as it would appear if cleaved by the proper restriction enzyme.

5 Use the top row in the table as an example and guide.

Name Date Class

MiniLab 13-1 Matching Restriction Enzymes to Cleavage Sites, *continued*

Analysis

1. Use the example provided in the data table to illustrate a single-stranded dangling end of DNA.

2. Record the DNA base sequence that must be present on a piece of viral DNA if these ends could "stick to" the dangling bases in the example shown in the data table.

3. Are restriction enzymes very specific as to where they cleave DNA? Explain your answer and give an example.

MiniLab 13-2

Using Numbers

Storing the Human Genome

It has been estimated that the human genome consists of three billion nitrogen base pairs. How much room would all the genetic information in a single cell take up if it were printed in a book the size of a typical novel?

Procedure

1. Use the data table below.
2. Select a random page from a novel.
3. Follow the directions in the table. Record your calculations in your data table.

Data Table

Directions	Letters and numbers
A. Count the number of characters (letters, punctuation marks, and spaces) across one entire line of your selected page.	
B. Count the number of lines on the page.	
C. Calculate the number of characters on the page. (Multiply A × B.)	
D. Let one nitrogen base equal one character. Knowing that DNA is made of nitrogen base pairs, divide C by 2.	
E. Record the number of pages in your novel.	
F. Calculate the number of base pairs in your novel. (Multiply E × D.)	
G. Calculate the number of books the size of your novel needed to hold the human genome. (Divide 3 billion by F.)	

Analysis

1. What changes could be taken to improve the accuracy of this activity at steps A–C?

2. What assumption is being made at step G?

MiniLab 13-2

Storing the Human Genome, *continued*

3. Explain the logic for step D.

4. a. How many books the size of your novel would be needed to store the human genome?

b. How many books the size of your novel would be needed to store a typical bacterial genome? Assume there are three million base pairs in the genome of a bacterium.

Name Date Class

Chapter 13

Modeling Recombinant DNA

Preparation

Problem

How can you model recombinant DNA technology?

Objectives

In this BioLab, you will:

- **Model** the process of preparing recombinant DNA.
- **Analyze** a model for preparation of recombinant DNA.

Materials

white paper
colored pencils, red and green
tape
scissors

Safety Precautions

Always wear goggles in the lab. Be careful with sharp objects.

Skill Handbook

Use the **Skill Handbook** if you need additional help with this lab.

Procedure

1. Cut a lengthwise strip of paper from a sheet of white paper into a rectangle about 3 cm by 28 cm. This strip represents a long sequence of DNA containing a particular gene that you wish to combine with a plasmid.
2. Cut another lengthwise strip of paper into a rectangle about 3 cm by 10 cm. When taped into a ring, this piece of paper will represent a bacterial plasmid.
3. Use your colored pencils to color the longer strip red and the shorter strip green.
4. Write the following DNA sequence once on the shorter strip of paper and two times about 5 cm apart on the longer strip of paper.

   -G-G-A-T-C-C-
   -C-C-T-A-G-G-

5. After coloring the shorter strip of paper and writing the sequence on it, tape the ends together.
6. Assume that a particular restriction enzyme is able to cleave DNA in a staggered way as illustrated here.

   -G   G-A-T-C-C-
   -C-C-T-A-G   G-

   Cut the longer strand of DNA in both places. You now have a cleaved foreign DNA fragment containing a gene that can be inserted into the plasmid.
7. Once the sequence containing the foreign gene has been cleaved, cut the plasmid in the same way.
8. Splice the foreign gene into the plasmid by taping the paper together where the sticky ends pair properly. The new plasmid represents recombinant DNA.

Name Date Class

Modeling Recombinant DNA, *continued*

Chapter 13

9. Use the data table at right. Relate the steps of producing recombinant DNA to the activities of the modeling procedure by explaining how the terms relate to the model.

Data Table

Term	BioLab model
Gene splicing	
Plasmid	
Restriction enzyme	
Sticky ends	
Recombinant DNA	

Analyze and Conclude

1. **Comparing and Contrasting** How does the paper model of a plasmid resemble a bacterial plasmid?

2. **Comparing and Contrasting** How is cutting with the scissors different from cleaving with a restriction enzyme?

3. **Thinking Critically** Enzymes that modify DNA, such as restriction enzymes, have been discovered and isolated from living cells. What functions do you think they have in living cells?

Name Date Class

MiniLab 14-1

Observing and Inferring

Marine Fossils

Certain sedimentary rocks are formed totally from the fossils of once-living ocean organisms called diatoms. The diatom fossils are often 1000 meters thick. These sedimentary rocks were at one time in the past under ocean water and were then lifted above sea level during periods of geological change.

Procedure

1. Prepare a wet mount of a small amount of diatomaceous earth. **CAUTION:** ***Use care in handling microscope slides and coverslips.***
2. Examine the material under low-power magnification.
3. Draw several of the different shapes you see.
4. Compare the shapes of the fossils you observe to present-day diatoms shown in the photograph. Remember, however, that the fossils you observe are probably only pieces of the whole organism.

Analysis

1. Describe the appearance of fossil diatoms.

2. How are fossil diatoms similar to and different from the diatoms in the photo? Can you use these similarities and differences to predict how diatoms have changed over time? Explain your answer.

3. What part of the original diatom did you observe under the microscope? How did this part survive millions of years? Why were the fossils you observed in pieces?

Name Date Class

MiniLab 14-2

Organizing Data

A Time Line

In this activity, you will construct a time line that is a scale model of the Geologic Time Scale. Use a scale in which 1 meter equals 1 billion years. Each millimeter then represents 1 million years.

Procedure

1. Use a meterstick to draw a continuous line down the middle of a 5 m strip of adding-machine tape.
2. At one end of the tape, draw a vertical line and label it "The Present."
3. Measure off the distance that represents 4.6 billion years ago. Draw a vertical line at that point and label it "Earth's Beginning."
4. Using the table below, plot the location of each event on your time line. Label the event and the number of years ago it occurred.

Geologic Time Scale

Event	Estimated years ago	Event	Estimated years ago
Earliest evidence of life	3.5 billion	First birds	150 million
Paleozoic era begins	540 million	Cretaceous period begins	146 million
First land plants	430 million	Dinosaurs become extinct	66 million
Mesozoic era begins	245 million	Cenozoic era begins	66 million
Triassic period begins	245 million	Primates appear	60 million
Jurassic period begins	208 million	Humans appear	200 000
First dinosaurs	225 million		

Analysis

1. Which era is the longest? The shortest?

2. In which eras did dinosaurs and mammals appear on Earth?

3. Which lived on Earth the longer time, dinosaurs or mammals?

Name Date Class

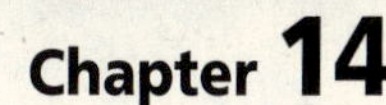

Determining a Fossil's Age

PREPARATION

Problem
How can you simulate radioactive half-life?

Objectives
In this BioLab you will:
- **Formulate models** Simulate the radioactive decay of K 40 into Ar 40 with pennies.
- **Collect data** Collect data to determine the amount of K 40 present after several half-lives.
- **Make and use a graph** Graph your data and use its values to determine the age of rocks.

Materials
shoe box with lid
100 pennies
graph paper

Skill Handbook
Use the **Skill Handbook** if you need additional help with this lab.

PROCEDURE

1. Use the data table below.
2. Place 100 pennies in a shoe box.
3. Arrange the pennies so that their "head" sides are facing up. Each "head" represents an atom of K 40, and each "tail" an atom of Ar 40.
4. Record the number of "heads" and "tails" present at the start of the experiment. Use the row marked "0" in the data table.
5. Cover the box. Then shake the box well. Let the shake represent one half-life of K 40, which is 1.3 billion years.

Data Table

Number of Shakes (half lives)	Number of Heads (K 40 atoms left)				
	Trial 1	Trial 2	Trial 3	Totals	Average
0					
1					
2					
3					
4					
5					

6. Remove the lid and record the number of "heads" you see facing up. Remove all the "tail" pennies.
7. To complete the first trial, repeat steps 5 and 6 four more times.
8. Run two more trials and determine an average for the number of "heads" present at each half-life.
9. Draw a full-page graph. Plot your average values on the graph. Plot the number of half-lives for K 40 on the *x*-axis and the number of "heads" on the *y*-axis. Connect the points with a line. Remember, each half-life mark on the graph axis for K 40 represents 1.3 billion years.

Chapter 14

Determining a Fossil's Age, *continued*

Analyze and Conclude

1. **Applying Concepts** What symbol represented an atom of K 40 in this experiment? What symbol represented an atom of Ar 40?

2. **Thinking Critically** Compare the numbers of protons and neutrons of K 40 and Ar 40. (Consult the Periodic Chart in the Appendix for help.) Can Ar 40 change back to K 40? Explain your answer, pointing out what procedural part of the experiment supports your answer.

3. **Defining Operationally** Define the term half-life. What procedural part of the simulation represented a half-life period of time in the experiment?

4. **Communicating** Explain how scientists use radioactive dating to approximate a fossil's age.

5. **Making and Using Graphs** You are attempting to determine the age of a rock sample. Use your graph to read the rock's age if it has:
   a. 70% of its original K 40 amount.

   b. 35% of its original K 40 amount.

   c. 10% of its original K 40 amount.

Name Date Class

MiniLab 15-1

Formulating Models

Camouflage Provides an Adaptive Advantage

Camouflage is a structural adaptation that allows organisms to blend with their surroundings. In this activity, you'll discover how natural selection can result in camouflage adaptations in organisms.

Procedure

1. Working with a partner, punch 100 dots from a sheet of white paper with a paper hole punch. Repeat with a sheet of black paper. These dots will represent black and white insects.
2. Scatter both white and black dots on a sheet of black paper.
3. Decide whether you or your partner will role-play a bird.
4. The "bird" looks away from the paper, then turns back, and immediately picks up the first dot he or she sees.
5. Repeat step 4 for one minute.

Analysis

1. What color dots were most often collected?

____

____

2. How does color affect the survival rate of insects?

____

____

____

3. What might happen over many generations to a similar population in nature?

____

____

____

MiniLab 15-2

Collecting Data

Detecting a Variation

Pick almost any trait—height, eye color, leaf width, or seed size—and you can observe how the trait varies in a population. Some variations are an advantage to an organism and some are not.

Procedure

1. Copy the data table shown here, but include the lengths in millimeters (numbers 25 through 45) that are missing from this table.

Data Table

Length in mm	20	21	22	23	24	–	46	47	48	49	50
Checks											
My Data—Number of shells											
Class Data—Number of shells											

2. Use a millimeter ruler to measure a peanut shell's length. In the Checks row, check the length you measured.
3. Repeat step 2 for 29 more shells.
4. Count the checks under the length and enter the total in the row marked My Data.
5. Use class totals to complete the row marked Class Data.

Analysis

1. Was there variation among the lengths of peanut shells? Use specific class totals to support your answer.

____

____

2. If larger peanut shells were a selective advantage, would this be stabilizing, directional, or disruptive selection? Explain your answer.

____

____

____

____

Natural Selection and Allelic Frequency

Chapter 15

PREPARATION

Problem

How does natural selection affect allelic frequency?

Objectives

In this BioLab, you will:

- **Simulate** natural selection by using beans of two different colors.
- **Calculate** allelic frequencies over five generations.
- **Demonstrate** how natural selection can affect allelic frequencies over time.
- **Use the Internet** to collect and compare data from other students.

Materials

colored pencils (2)
paper bag
graph paper
pinto beans
white navy beans

Skill Handbook

Use the **Skill Handbook** if you need additional help with this lab.

PROCEDURE

1. Use the data table.
2. Place 50 pinto beans and 50 white navy beans into the paper bag.
3. Shake the bag. Remove two beans. These represent one rabbit's genotype. Set the pair aside, and continue to remove 49 more pairs.
4. Arrange the beans on a flat surface in two columns representing the two possible rabbit phenotypes, gray (genotypes *GG* or *Gg*) and white (genotype *gg*).
5. Examine your columns. Remove 25 percent of the gray rabbits and 100 percent of the white rabbits. These numbers represent a random selection pressure on your rabbit population. If the number you calculate is a fraction, remove a whole rabbit to make whole numbers.
6. Count the number of pinto and navy beans remaining. Record this number in your data table.
7. Calculate the allelic frequencies by dividing the number of beans of one type by 100. Record these numbers in your data table.
8. Begin the next generation by placing 100 beans into the bag. The proportions of pinto and navy beans should be the same as the percentages you calculated in step 7.
9. Repeat steps 3 through 8, collecting data for five generations.
10. **Post your data** on the Glencoe Science Web Site at **www.glencoe.com/sec/science.**
11. Graph the frequencies of each allele over five generations. Plot the frequency of the allele on the vertical axis and the number of the generation on the horizontal axis. Use a different colored pencil for each allele.

Natural Selection and Allelic Frequency, *continued*

Data Table

	Allele G			Allele *g*		
Generation	**Number**	**Percentage**	**Frequency**	**Number**	**Percentage**	**Frequency**
Start	50	50	0.50	50	50	0.50
1						
2						
3						
4						
5						

Analyze and Conclude

1. **Analyzing Data** Did either allele disappear? Why or why not?

2. **Thinking Critically** What does your graph show about allelic frequencies and natural selection?

3. **Making Inferences** What would happen to the allelic frequencies if the number of eagles declined?

4. **Interpreting Data** Explain any differences in allelic frequencies you observed between your data and the data from the Internet.

Name Date Class

MiniLab 16-1

Comparing and Contrasting

Comparing Old and New World Monkeys

In this activity, you will gather and then compare data about Old World monkeys and New World monkeys.

Procedure

1. Use the data table below.
2. Examine the diagrams on page 433 of your text.
3. Complete the data table.

Analysis

1. Why would a low body weight be helpful for an animal with an arboreal life style?

2. Which group appears to be more closely related to humans? Explain your answer. (Note: Humans have eight premolars, twelve molars, and a total of 32 teeth.)

Data Table

Characteristic	New World monkey	Old World monkey
Number of premolars ($\frac{1}{4}$ jaw)		
Number of molars ($\frac{1}{4}$ jaw)		
Total teeth in mouth		
Nostril position		
Tail		
Body weight	0.14 to 11 kg	1.2 to 30 kg

MiniLab 16-2

Analyzing Information

Compare Human Proteins with Those of Other Primates

Scientists use differences in amino acid sequences in proteins to determine evolutionary relationships of living species. In this activity, you'll compare representative short sequences of amino acids of a protein among other groups of primates to determine their evolutionary history.

Procedure

1. Use the data table below.
2. For each primate listed in the table at right, determine how many amino acids differ from the human sequence. Record these numbers in the data table.
3. Calculate the percentage differences by dividing the numbers by 15 and multiplying by 100. Record the numbers in your data table.

Table 16.1 Amino Acid Sequences in Primates

Baboon	Chimp	Lemur	Gorilla	Human
ASN	SER	ALA	SER	SER
THR	THR	THR	THR	THR
THR	ALA	SER	ALA	ALA
GLY	GLY	GLY	GLY	GLY
ASP	ASP	GLU	ASP	ASP
GLU	GLU	LYS	GLU	GLU
VAL	VAL	VAL	VAL	VAL
ASP	GLU	GLU	GLU	GLU
ASP	ASP	ASP	ASP	ASP
SER	THR	SER	THR	THR
PRO	PRO	PRO	PRO	PRO
GLY	GLY	GLY	GLY	GLY
GLY	GLY	SER	GLY	GLY
ASN	ALA	HIS	ALA	ALA
ASN	ASN	ASN	ASN	ASN

Analysis

1. Which primate is most closely related to humans? Least closely related?

____

____

2. On another sheet of paper, construct a diagram of primate evolutionary relationships that most closely fits your results.

Data Table

Primate	Amino acids different from humans	Percent difference
Baboon		
Chimpanzee		
Gorilla		
Lemur		

Name Date Class

Chapter 16

Comparing Skulls of Three Primates

PREPARATION

Problem

How do skulls of primates provide evidence for human evolution?

Objectives

In this BioLab, you will:

- **Determine** how paleoanthropologists study early human ancestors.
- **Compare and contrast** the skulls of australopithecines, gorillas, and modern humans.

Materials

metric ruler
protractor
copy of skull diagrams

Skill Handbook

Use the **Skill Handbook** if you need additional help with this lab.

PROCEDURE

1. Your teacher will provide copies (1/2 natural size) of the skulls of *Australopithecus africanus*, *Gorilla gorilla*, and *Homo sapiens*.
2. The rectangles drawn over the skulls represent the areas of the brain (upper rectangle) and face (lower rectangle). On each skull, determine and record the area of each rectangle (length × width).
3. Measure the diameters of the circles in each skull. Multiply these numbers by 200 cm^2. The result is the cranial capacity (brain volume) in cubic centimeters.
4. The two heavy lines projected on the skulls are used to measure how far forward the jaw protrudes. Use your protractor to measure the outside angle (toward the right) formed by the two lines.
5. Complete the data table.

Data Table

	Gorilla	Anstralopithecus	Modern human
1. Face area in cm^2			
2. Brain area in cm^2			
3. Is brain area smaller or larger than face area?			
4. Is brain area 3 times larger than face area?			
5. Cranial capacity in cm^3			
6. Jaw angle			
7. Does lower jaw stick out in front of nose?			
8. Is sagittal crest present?			
9. Is browridge present?			

Chapter 16

Comparing Skulls of Three Primates, *continued*

Analyze and Conclude

1. **Comparing and Contrasting** How would you describe the similarities and differences in face-to-brain area in the three primates?

2. **Interpreting Observations** How do the cranial capacities compare among the three skulls? How do the jaw angles compare?

3. **Drawing Conclusions** Based on your findings, what statements can you make about the placement of australopithecines in human evolution?

Primate Skulls for Biolab, Chapter 16

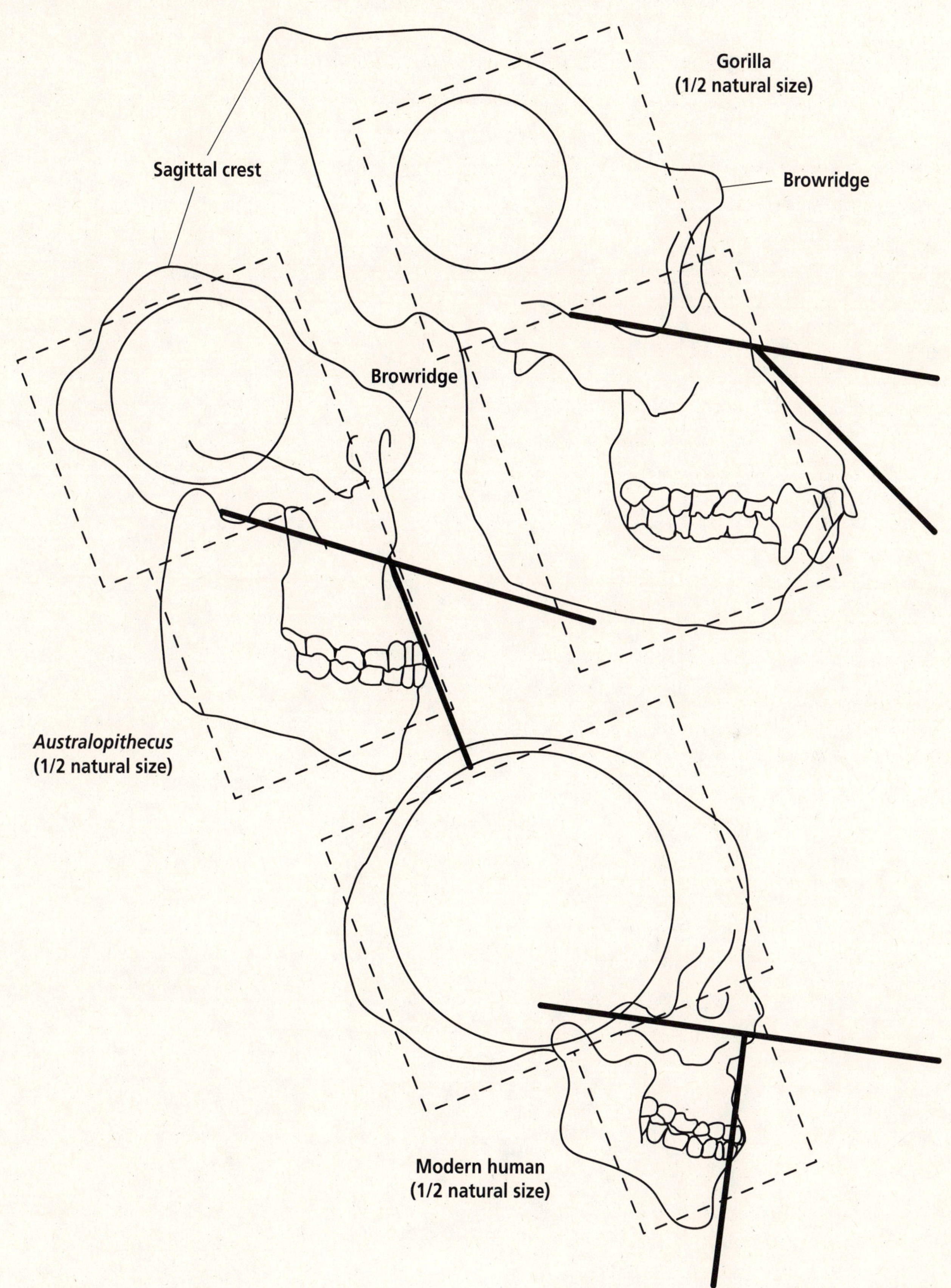

MiniLab 17-1 Using a Dichotomous Key

Classifying

How could you identify a tree growing in front of your school? You might ask a local expert, or you could use a manual or field guide that contains descriptive information and keys about trees. A key is a set of descriptive sentences that is subdivided into steps. A dichotomous key has two descriptions at each step. You follow the steps until the key reveals the name of the tree.

Procedure

1. Using a few leaves from local trees and a dichotomous key for trees of your area, identify the tree from which each leaf came. To use the key, study one leaf. Then choose the one statement from the first pair that most accurately describes the leaf. Continue following the key until you identify the leaf's tree. Repeat the process for each leaf.
2. Glue each leaf on a separate sheet of paper. For each leaf, record the tree's name.

Analysis

1. What is the function of a dichotomous key?

___

___

2. List three different characteristics used in your key.

___

___

3. As you used the key, did the characteristics become more general or more specific?

___

Name Date Class

MiniLab 17-2

Classifying

Using a Cladogram to Show Relationships

Cladograms were developed by Willie Hennig. They use derived characteristics to illustrate evolutionary relationships.

Procedure

1. The following table shows the presence or absence of six derived traits in the seven dinosaurs that are labeled A–G.
2. Use the information listed in the table to answer the questions below.

Derived traits of dinosaurs

Dinosaur trait	A	B	C	D	E	F	G
Hole in hip socket	yes	yes	yes	yes	yes	yes	yes
Extension of pubis bone	no	no	no	yes	yes	yes	yes
Unequal enamel on teeth	no	no	no	no	yes	yes	yes
Skull has "shelf" in back	no	no	no	no	no	yes	yes
Grasping hand	yes	yes	yes	no	no	no	no
Three-toed hind foot	yes	yes	no	no	no	no	no

Analysis

1. Copy the partially completed cladogram on page 467 of your text. Complete the missing information on the right side.

____

2. How many traits does dinosaur F share with dinosaur C, with dinosaur D, and with dinosaur E?

____

3. Dinosaurs A and B form a grouping called a clade. Dinosaurs A, B, and C form another clade. What derived trait is shared only by the A and B clade? By the A, B, and C clade? By the D, E, F, and G clade?

____

____

4. Traits that evolved early, such as the hole in the hip socket, are called primitive traits. Traits that evolved later, such as a grasping hand, are called derived traits. Are primitive traits typical of broader or smaller clades? Are derived traits typical of broader or smaller clades? Give an example in each case.

____

____

Name Date Class

Chapter 17

Making a Dichotomous Key

Preparation

Problem
How is a dichotomous key made?

Objectives
In this BioLab, you will:
- **Classify** organisms on the basis of structural characteristics.
- **Develop** a dichotomous key.

Materials
sample keys from guidebooks
metric ruler

Skill Handbook
Use the **Skill Handbook** if you need additional help with this lab.

Procedure

1. Study the drawings of beetles on page 475 of your text.
2. Choose one characteristic of the beetles and classify the beetles into two groups based on that characteristic. Take measurements if you wish.
3. Record the chosen characteristic in a diagram like the one shown. Write the numbers of the beetles in each group on your diagram.
4. Continue to form subgroups within your two groups based on different characteristics. Record the characteristics and numbers of the beetles in your diagram until you have only one beetle in each group.
5. Using the diagram you have just made, make a dichotomous key for the beetles. Remember that each numbered step should contain two choices for classification. Begin with 1A and 1B. For help, examine sample keys provided by your teacher.
6. Exchange dichotomous keys with another team. Use their keys to identify the beetles.

Analyze and Conclude

1. **Comparing and Contrasting** Was the dichotomous key you constructed exactly like those of the other students? Explain your answer.

____

____

____

____

____

INVESTIGATE BioLab

Making a Dichotomous Key, *continued*

2. Analyzing Data What characteristics were most useful for making a classification key for beetles? What characteristics were not useful?

3. Thinking Critically Why do keys typically offer only two choices and not more?

Name Date Class

MiniLab 18-1

Measuring in SI

Measuring a Virus

Can you use a light microscope to view a virus? Find out by measuring the size of a poliovirus in the photo on page 490 of your text and then comparing it to 0.2 µm, the size limit for viewing objects with a light microscope.

Procedure

1. Use the data table below.

Data Table

Values to measure and calculate	Measurement
Length of photo line in mm	
Diameter of poliovirus in mm	
Diameter of poliovirus in µm	

2. Examine the photo on page 490 of your text showing many polioviruses. The horizontal line you see would measure only 0.4 micrometer (µm) in length if the photo was not magnified 172 500×. Use this line for reference.
3. Calculate the diameter of one poliovirus. First, measure the length of the reference line in millimeters. Record the value in the table. Then, measure the diameter of a poliovirus in millimeters. Record the value in the table.
4. Use the following equation to calculate the actual diameter of the poliovirus (X). Record your answer in the table.

$$\frac{\text{photo line length in mm } (A)}{\text{diameter of virus in mm } (B)} = \frac{0.4\ \mu\text{m}}{\text{diameter of virus in } \mu\text{m } (X)}$$

Analysis

1. Explain why you cannot see viruses with a light microscope. Use specific numbers in your answer.

____

____

2. A bacterial cell may be 10 µm in size. How many polioviruses could fit across the top of such a bacterium?

____

____

Name Date Class

MiniLab 18-2

Observing

Bacteria Have Different Shapes

Bacteria come in three different shapes: spherical (coccus), rodlike (bacillus), and spiral shaped (spirillum). They may appear singly or in pairs, chains, or clusters. Each species has a typical shape and reaction to Gram stain.

Procedure

1. Obtain slides of bacteria from your teacher.
2. Using low power, locate bacteria of one shape. Switch to high power. Look for individual cells and observe their shape. Observe also the size of the cells and their color. Then look for groups of bacterial cells to determine their arrangement. **CAUTION:** ***Use caution when working with a microscope and microscope slides.***
3. Repeat step 2 for bacteria with the other shapes. Then, compare the sizes of the bacteria.
4. Draw a diagram of each type of bacteria.

Analysis

1. How do the sizes of the three bacteria compare?

2. Which of the bacteria were Gram negative?

3. What adaptive advantage might there be for bacteria to form groups of cells?

Name Date Class

How sensitive are bacteria to antibiotics?

Chapter 18

PREPARATION

Problem

How can you determine which antibiotic most effectively kills specific bacteria?

Hypotheses

Decide on one hypothesis that you will test. Your hypothesis might be that the antibiotic with the widest zone of inhibition most effectively kills bacteria.

Objectives

In this BioLab, you will:

- **Compare** how effectively different antibiotics kill specific bacteria.
- **Determine** the most effective antibiotic to tread an infection that these bacteria might cause.

Possible Materials

cultures of bacteria
sterile nutrient agar petri dishes
antibiotic disks
sterile disks of blank filter paper
marking pen
long-handled cotton swabs
forceps
37°C incubator
metric ruler

Safety Precautions

Always wear goggles in the lab. Although the bacteria you will work with are not disease-causing, be careful not to spill them. Wash your hands with soap immediately after handling any bacterial culture. Carefully clean your work area after you finish. Follow your teacher's instructions about disposal of your swabs, cultures, and petri dishes.

Skill Handbook

Use the **Skill Handbook** if you need additional help with this lab.

DESIGN YOUR OWN BioLab

How sensitive are bacteria to antibiotics?, *continued*

PLAN THE EXPERIMENT

1. Examine the materials provided by your teacher, and study the two photos. As a group, make a list of ways you might investigate your hypothesis.
2. Agree on one way that your group could investigate your hypothesis. Design an experiment in which you can collect quantitative data.
3. Make a list of numbered directions. In your list, include the amounts of each material you will need. If possible, use no more than one petri dish per person.
4. Design and construct a table for recording data. To do this, carefully consider what data you need to record and how you will measure the data. For example, how will you measure what happens around the antibiotic disks as the antibiotic diffuses into the agar?

Check the Plan

Discuss the following points with other group members to decide on your final procedure.

1. How you will set up your petri dishes. How many antibiotics can you test on one petri dish? How will you measure the effectiveness of each antibiotic? What will be your control?
2. Will you add the bacteria or the antibiotic disks first?
3. What will you do to prevent other bacteria from contaminating the petri dishes?
4. How often will you observe the petri dishes?
5. ***Make sure your teacher has approved your experimental plan before you proceed.***
6. Carry out your experiment. **CAUTION:** ***Wash your hands with soap and water after handling dishes of bacteria.***

ANALYZE AND CONCLUDE

1. **Measuring in SI** How did you measure the zones of inhibition? Why did you do it this way?

___

___

2. **Drawing Conclusions** Suppose you were a physician treating a patient infected with these bacteria. Which antibiotic would you use? Why?

___

___

3. **Analyzing the Procedure** What limitations does this technique have? If these bacteria were infecting a person, what other tests might increase your confidence about treating the person with the antibiotic that appears most effective against these bacteria?

___

___

___

Name Date Class

MiniLab 19-1

Observing and Inferring

Observing Ciliate Motion

The cilia on the surface of a paramecium move so that the cell normally swims through the water with one end directed forward. But when this end bumps into an obstacle, the paramecium responds by changing direction.

Procedure

1. Observe a *Paramecium* culture that has had boiled, crushed wheat seeds in it for several days.
2. Carefully place a drop of water containing wheat seed particles on a microscope slide. Gently add a coverslip.
3. Using low power, locate a paramecium near some wheat seed particles. **CAUTION:** ***Use caution when working with a microscope, glass slides, and coverslips.***
4. Watch the paramecium as it swims around among the particles. Record your observations of the organism's responses each time it contacts a particle.

Analysis

1. Describe what a paramecium does when it encounters an obstacle.

2. How long does the paramecium's response last?

3. Describe any changes in the shape of the paramecium as it moved among the particles.

Name Date Class

MiniLab 19-2 Going on an Algae Hunt

Observing

Pond water may be teeming with organisms. Some are macroscopic organisms, but the majority are microscopic. Some may be heterotrophs, and others autotrophs. How can you tell them apart?

Procedure

1. Use the data table below.

Data Table

Diagram	Motile/Nonmotile	Unicellular/Multicellular

2. Place a drop of pond water onto a glass slide and add a coverslip. **CAUTION:** ***Use caution when working with a microscope, glass slides, and coverslips.***
3. Observe the pond water under low magnification of your microscope, and look for algae that may be present. Algae from a pond will usually be green or yellow-green in color.
4. Diagram several different species of algae in your data table and indicate if each is motile or nonmotile. Indicate if the algae are unicellular or multicellular.

Analysis

1. What characteristic distinguished algae from any protozoans that may have been present?

____

2. Explain how the characteristic in question 1 categorizes algae as autotrophs.

____

3. Did you observe any relationship between movement and size? Explain your answer.

____

Chapter 19

How do *Paramecium* and *Euglena* respond to light?

PREPARATION

Problem

Do both *Paramecium* and *Euglena* respond to light and do they respond in different ways? Among your group, decide on one type of protist activity that would constitute a response to light.

Hypotheses

Decide on one hypothesis that you will test. Your hypothesis might be that *Paramecium* will not respond to light and *Euglena* will respond, or that *Paramecium* will move away from light and *Euglena* will move toward light.

Objectives

In this BioLab, you will:

- **Prepare** slides of *Paramecium* and *Euglena* cultures and observe swimming patterns in the two organisms.
- **Compare** how these two different protists respond to light.

Possible Materials

Euglena culture
Paramecium culture
microscope
microscope slides
dropper
methyl cellulose
coverslips
metric ruler
index cards
scissors
toothpicks

Safety Precautions

Always wear goggles in the lab. Use caution when working with a microscope, glass slides, and coverslips. Wash your hands with soap and water immediately after working with protists and chemicals.

Skill Handbook

Use the **Skill Handbook** if you need additional help with this lab.

PLAN THE EXPERIMENT

1. Decide on an experimental procedure that you can use to test your hypothesis.
2. Record your procedure, step-by-step, and list the materials you will be using.
3. Design a data table in which to record observations and results.

Check the Plan

Discuss the following points with other group members to determine your final procedure.

1. What variables will you have to measure?
2. What will be your control?
3. What will be the shape of the light-controlled area(s) on your microscope slide?
4. Decide who will prepare materials, make observations, and record data.
5. ***Make sure your teacher has approved your experimental plan before you proceed further.***
6. To mount drops of *Paramecium* culture and *Euglena* culture on microscope slides, use a toothpick to place a small ring of methyl cellulose on a clean microscope slide. Place a drop of *Paramecium* or *Euglena* culture within this ring. Place a coverslip over the ring and culture. The thick consistency of methyl cellulose should slow down the organisms for easy observation.
7. Make preliminary observations of swimming *Paramecia* and *Euglena*. Then think again about the observation times that you have planned. Maybe you will decide to allow more or less time between your observations.
8. Carry out your experiment.

DESIGN YOUR OWN BioLab

How do *Paramecium* and *Euglena* respond to light?, *continued*

Chapter 19

Analyze and Conclude

1. **Checking Your Hypothesis** Did your data support your hypothesis? Why or why not?

2. **Comparing and Contrasting** Compare and contrast the responses of *Paramecium* and *Euglena* to light and darkness. What explanations can you suggest for their behavior?

3. **Making Inferences** Can you use your results to suggest what sort of responses to light and darkness you might observe using other heterotrophic or autotrophic protists?

Name Date Class

MiniLab 20-1

Observing and Inferring

Growing Mold Spores

Any mold spore that lands in a favorable place can germinate and produce hyphae. Can you identify a condition necessary for the growth of bread mold spores?

Procedure

1. Place two slices of freshly baked bakery bread on a plate. Sprinkle some water on one slice to moisten its surface. Leave both slices uncovered for several hours.
2. Sprinkle a little more water on the moistened slice, and place both slices in their own plastic, self-seal bags. Trap air in each bag so that the plastic does not touch the bread's surface. Then seal the bags and place them in a darkened area at room temperature.
3. After five days, remove the bags and look for mold.
4. Remove a small piece of mold with a forceps, place it on a slide in a drop of water, and add a coverslip. Observe the mold under a microscope's low power and high power. **CAUTION:** ***Use caution when working with a microscope, glass slides, and coverslips. Wash your hands with soap and water after working with mold. Dispose of the mold as your teacher directs.***

Analysis

1. Did you observe mold growth on the moistened bread? On the dry bread? How does this experiment demonstrate that there are mold spores in your classroom?

2. What conclusions can you draw about the conditions necessary for the growth of a bread mold?

Name Date Class

Classifying

MiniLab 20-2 Examining Mushroom Gills

Spore prints can often help in mushroom identification by revealing the pattern of a mushroom's gills and the color of its spores. Use this technique to see how a mushroom's gills are arranged.

Procedure

1. Break off the stalks from several grocery-store mushrooms. Place the caps in a paper bag for a few days.
2. When the undersides of the caps are very dark brown, set the caps, gill side down, on a white sheet of paper. Be sure that the gills are touching the surface of the paper.
3. After leaving the caps undisturbed overnight, carefully lift the caps from the paper and observe the results.
4. Wash your hands with soap and water. Dispose of fungi as your teacher directs.

Analysis

1. What color are the spores on the paper?

____

2. How does the pattern of spores on the paper compare with the arrangement of gills on the underside of the mushroom cap that produced it?

____

____

Name Date Class

Chapter 20

Does temperature affect the metabolic activity of yeast?

PREPARATION

Problem

How can you determine the affect of temperature on the metabolism of yeast? Brainstorm ideas among the members of your group.

Hypotheses

Decide on one hypothesis that you will test. Your hypothesis might be that low temperature slows down the metabolic activity of yeast, or that high temperature speeds up the metabolic activity of yeast.

Objectives

In this BioLab, you will:

- **Measure** the rate of yeast metabolism using a BTB color change as a rate indicator.
- **Compare** the rates of yeast metabolism at several temperatures.
- **Use the Internet** to collect and compare data from other students.

Possible Materials

bromothymol blue solution (BTB)
straw
small test tubes (4)
large test tubes (3)
one-hole stoppers with glass tube inserts for large test tubes (3)
yeast/white corn syrup mixture
water/white corn syrup mixture
water/yeast mixture
test-tube rack
250 mL beakers (3)
ice cubes
Celsius thermometer
hot plate
50 mL graduated cylinder
glass-marking pencil
10 cm rubber tubing (3)
aluminum foil

Safety Precautions

Always wear goggles in the lab. Be careful in attaching rubber tubing to the glass tube inserts in the stoppers. Avoid touching the top of the hot plate. Wash your hands thoroughly after cleaning out test tubes at the end of your experiments.

Skill Handbook

Use the **Skill Handbook** if you need additional

Does temperature affect the metabolic activity of yeast?, *continued*

Chapter 20

Plan the Experiment

1. Decide on ways to test your group's hypothesis.
2. Record your procedure, and list the materials and amounts of solutions that you will use.
3. Design a data table for recording your observations.
4. Pour 5 mL of BTB solution into a test tube. Use a straw to blow gently into the tube until you see a series of color changes. Cover this tube with aluminum foil, and set it aside in a test-tube rack. Record your observations of the color changes caused by carbon dioxide in your breath.

Check the Plan

Discuss the following to decide on your procedure.

1. What data on color change and time will you collect? How will you record your data?
2. What variables will you control?
3. What control will you use?
4. Assign tasks for each member of your group.
5. ***Make sure your teacher has approved your experimental plan before you proceed further.***
6. Carry out your experiment.
7. Visit the Glencoe Science Web Site at **www.glencoe.com/sec/science** to **post your data.**

Analyze and Conclude

1. **Checking Your Hypotheses** Explain whether your data support your hypothesis. Use your experimental data to support or reject your hypothesis concerning temperature effects on the rate of yeast metabolism.

2. **Using the Internet** Did the data from the Internet support your hypothesis? Explain your answer.

3. **Making Inferences** What must be the role of white corn syrup in this experiment?

4. **Identifying Variables** Describe some variables that had to be controlled in this experiment. Explain how you controlled each variable.

5. **Drawing Conclusions** Describe the control used in your experiment and how your experimental results enabled you to draw conclusions about the effect of temperature on yeast metabolism. Did your experiment clearly show that differences in rates of yeast metabolism were due to temperature differences?

Name Date Class

MiniLab 21-1

Applying Concepts

Examining Land Plants

Liverworts are considered to be one of the simplest of all land plants. They show many of the adaptations that other land plants have evolved that enable them to survive on a land environment.

Procedure

1. Examine a living or preserved sample of Marchantia. **CAUTION:** ***Wear disposable latex gloves when handling preserved materials.***
2. Note and record the following observations. Is the plant unicellular or multicellular? Does it have a top and bottom? How do these differ? Is it one cell in thickness or many cells thick? Does the plant seem to grow upright like a tree or is it flat to the ground?
3. Use a dissecting microscope to examine its top and bottom surfaces. Are tiny holes or pores present? If you answer "yes," which surface has pores?

Analysis

1. How might having a multicellular, thick body be an advantage to life on land?

2. Are any structures present that resemble roots? How might they help a land plant?

3. How might its growth pattern help it survive land?

4. What might be the role of any pores observed on the plant? Why is the location of the pores critical to surviving on a land environment?

MiniLab 21-2

Comparing and Contrasting

Looking at Modern and Fossil Plants

Many modern-day plants have relatives that are known only from the fossil record. Are modern-day plants similar to their fossil relatives? Are there any differences?

Procedure

1. Examine a preserved or living sample of *Lycopodium*, a club moss. **CAUTION:** ***Wear disposable latex gloves when handling preserved material.***
2. Note and record the following observations:
   a. Does the plant grow flat or upright like a tree?
   b. Describe the appearance of its leaves and its stem.
   c. Measure the plant's height and diameter in centimeters.
3. Diagram A on page 586 of your text is a representation of a fossil relative called *Lepidodendron*. Record the same observations (a–c).
4. Repeat steps a–c only this time use a preserved or living sample of *Equisetum*, a horsetail. Compare it to Diagram B on page 586 of your text, a representation of a fossil relative called *Calamites*.

Analysis

1. Describe the similarities and differences between *Lycopodium* and *Lepidodendron*. Do your observations justify their closeness as relatives? Explain.

2. Describe the similarities and differences between *Equisetum* and *Calamites*. Do your observations justify their closeness as relatives? Explain.

Name Date Class

Researching Trees on the Internet

Chapter 21

PREPARATION

Problem

Use the Internet to find different trees that would be suitable for planting in your community.

Objectives

In this BioLab, you will:

- **Research** the characteristics of five different trees.
- **Use the Internet** to collect and compare data from other students.
- **Conclude** which trees would be most suitable for planting in your community.

Materials

Internet access

Skill Handbook

Use the **Skill Handbook** if you need additional help with this lab.

PROCEDURE

1. Use the data table.
2. Pick five trees that you wish to research. Note: Your teacher may provide you with suggestions if necessary.
3. Go to the Glencoe Science Web Site at **www.glencoe.com/sec/science** to find links to information needed for this BioLab.
4. Record the information in your data table.

Data Table

	1	2	3	4	5
Tree Name (common name)					
Scientific Name					
Kingdom					
Division					
Soil/Water Preference					
Temperature Tolerance					
Height at Maturity					
Speed of Growth					
General Shape					
Disease Resistance/Problems					
Special Care					
Leaf Shape					
Shade Provider					
Additional Information					

Name Date Class

Chapter 21

Researching Trees on the Internet, *continued*

Analyze and Conclude

1. **Defining Operationally** Explain the difference between trees classified as either Coniferophyta or Anthophyta.

2. **Analyzing** Was the information provided on the Internet helpful in completing your data table? Explain your answer.

3. **Thinking Critically** What do you consider to be the most important characteristic when deciding on the most suitable tree for your community? Explain your answer.

4. **Using the Internet** Using the information you gathered from the Internet, which tree species would most likely be the:
   a. most suitable for your community? Explain your answer.

   b. least suitable for your community? Explain your answer.

5. **Applying** Explain why tree selections would differ if your community were located in:
   a. Tucson, Arizona

   b. Los Angeles, California

   c. Fairbanks, Alaska

Name Date Class

Experimenting

MiniLab 22-1 Identifying Fern Sporangia

When you admire a fern growing in a garden or forest, you are admiring the plant's sporophyte generation. Upon further examination, you should be able to see evidence of spores being formed. Typically, the evidence you are looking for can be found on the underside of the ferns' fronds.

Procedure

1. Place a drop of water and a drop of glycerin at opposite ends of a glass slide.
2. Use forceps to gently pick off one sorus from a frond. Place it in the drop of water and add a coverslip.
3. Add a second sorus to the glycerin and add a coverslip.
4. Observe both preparations under low-power magnification and note any similarities and differences. Look for large sporangia (resembling heads on a stalk) and spores (tiny round bodies released from a sporangium). **CAUTION:** ***Use caution when working with a microscope, microscope slides, and coverslips.***

Analysis

1. Were spores more visible in water or in glycerin?

____

2. What did the glycerin do to the sporangium?

____

____

3. Formulate a hypothesis that may explain how sporangia naturally burst.

____

____

4. Formulate a hypothesis that may explain how sporangia were affected by glycerin.

____

____

MiniLab 22-2

Comparing and Contrasting

Comparing Seed Types

Anthophytes are classified into two classes, the monocotyledons (monocots) and dicotyledons (dicots) based on the number of seed leaves.

Procedure

1. Use the data table below.
2. Examine the variety of seeds given to you. Use forceps to gently remove the seed coat or covering from each seed if one is present.
3. Determine the number of cotyledons present. If two cotyledons are present, the seed will easily separate into two equal halves. If one cotyledon is present, it will not separate into halves. Record your observations in the data table.
4. Add a drop of iodine stain to rice and a lima bean seed. Note the color change. **CAUTION:** ***Wash your hands with soap and water after handling chemicals.*** Record your observations in the data table.

Data Table

Seed name	Number of cotyledons	Monocot or dicot	Color with iodine
Lima bean			
Rice			
Pea			—
Rye			—

Analysis

1. Starch turns purple when iodine is added to it. Describe the color change when iodine was added to rice and lima bean seeds.

2. Hypothesize why seeds contain stored starch.

Name Date Class

How can you make a key for identifying conifers?

Chapter 22

Preparation

Problem

What kinds of characteristics can be used to create a key for identifying different kinds of conifers?

Hypotheses

State your hypothesis according to the kinds of characteristics you think will best serve to distinguish among several conifer groups. Explain your reasoning.

Objectives

In this BioLab, you will:

- **Compare** structures of several different conifer specimens.
- **Identify** which characteristics can be used to distinguish one conifer from another.
- **Communicate** to others the distinguishing features of different conifers.

Possible Materials

twigs, branches, and cones from several different conifers that have been identified for you

Safety Precautions

Always wash your hands after handling biological materials. Always wear goggles in the lab.

Skill Handbook

Use the **Skill Handbook** if you need additional help with this lab.

Plan the Experiment

1. Make a list of characteristics that could be included in your key. You might consider using shape, color, size, habitat, or other factors.
2. Determine which of those characteristics would be most helpful in classifying your conifers.
3. Determine in what order the characteristics should appear in your key.
4. Decide how to describe each characteristic.

Check the Plan

1. The traits described at each fork in a key are often pairs of contrasting characteristics. For example, the first fork in a key to conifers might compare "needles grouped in bundles" with "needles attached singly."
2. Someone who is not familiar with conifer identification should be able to use your key to correctly identify any conifer it includes.
3. ***Make sure your teacher has approved your experimental plan before you proceed further.***
4. Carry out your plan by creating your key.

Name Date Class

How can you make a key for identifying conifers?, *continued*

Chapter 22

Analyze and Conclude

1. **Checking Your Hypothesis** Have someone outside your lab group try using your key to identify your conifer specimens. If they are unable to make it work, try to determine where the problem is and make improvements.

2. **Making Inferences** Is there only one correct way to design a key for your specimens? Explain why or why not.

3. **Relating Concepts** Give one or more examples of situations in which a key would be a useful tool.

MiniLab 23-1

Observing

Examining Plant Tissues

Pipes are hollow. Their shape or structure allows them to be used efficiently in transporting water. Plant vascular tissues have this same efficiency in structure.

Procedure

1. Snap a celery stalk in half and remove a small section of "stringy tissue" from its inside.
2. Place the material on a glass slide. Add several drops of water. Place a second glass slide on top. **CAUTION:** ***Use caution when working with a microscope and slides.***
3. Press down evenly on the top glass slide with your thumb directly over the plant material.
4. Remove the top glass slide. Add more water if needed. Add a coverslip.
5. Examine the celery material under low- and high-power magnification. Diagram what you see.
6. Repeat steps 2–5 using some of the soft tissue inside the celery stalk.

Analysis

1. Describe the appearance of the stringy tissue inside the celery stalk. What may be the function of this tissue?

____

____

____

2. Describe the appearance of the soft tissue inside the celery stalk. What may be the function of this tissue?

____

____

____

3. Does the structure of these tissues suggest their functions?

____

____

Name Date Class

MiniLab 23-2

Observing

Looking at Stomata

The lens-shaped openings in the epidermis of a leaf allow gas exchange and help control water loss.

Procedure

1. Make a wet mount by tearing a leaf at an angle to expose a thin section of epidermis. Use tap water to make wet mounts of both the upper and lower epidermis.
2. Examine each of your slide preparations under the microscope. Draw or take down a written description of what you see. **CAUTION:** ***Use caution when working with a microscope, microscope slides, and coverslips.***
3. Make another wet mount using a 5 percent salt solution instead of tap water. Examine the slide under the microscope and record your observations.

Analysis

1. What do the cells of the leaf epidermis look like? What is their function?

____

____

2. How do the epidermal cells differ from guard cells? Which cells, if any, contain chloroplasts?

____

____

____

____

3. What differences in the stomata did you notice when you used a salt solution to prepare your wet mount? Can you explain what happened in terms of osmosis?

____

____

____

____

____

Name Date Class

INVESTIGATE BioLab

Determining the Number of Stomata on a Leaf

Chapter 23

PREPARATION

Problem

How can you count the total number of stomata on a leaf?

Objectives

In this BioLab, you will:

- **Measure** the area of a leaf.
- **Observe** the number of stomata seen under a high-power field of view.
- **Calculate** the total number of stomata on a leaf.

Materials

microscope
glass slide
water and dropper
green leaf from an onion plant
single-edged razor blade
ruler
glass cover

Safety Precautions

Wear latex gloves when handling an onion.

Skill Handbook

Use the **Skill Handbook** if you need additional help with this lab.

PROCEDURE

1. Use Data Tables 1 and 2.
2. Obtain an onion leaf and carefully cut it open lengthwise using a single-edged razor blade. **CAUTION:** ***Be careful when cutting with a razor blade.***
3. Measure the length and width of your onion leaf in millimeters. Record these values in Data Table 2.
4. Remove a small section of leaf and place it on a glass slide with the dark green side facing DOWN.
5. Add several drops of water and gently scrape away all green leaf tissue using a back and forth motion with the razor blade. An almost transparent layer of leaf epidermis will be left on the slide.
6. Add water and a cover glass to the epidermis and observe under low-power magnification. Locate an area where guard cells and stomata can clearly be seen. **CAUTION:** ***Use caution when working with a microscope, microscope slides, and coverslips.***
7. Switch to high-power magnification.
8. Count and record the number of stomata in your field of view. Consider this trial 1. Record your count in Data Table 1.

Data Table 1

Trial	Number of stomata
1	
2	
3	
4	
5	
Total	
Average	

9. Move the slide to a different area. Count and record the number of stomata in this field of view. Consider this trial 2.

Determining the Number of Stomata on a Leaf, *continued*

Chapter 23

10. Repeat step 9 three more times. Calculate the average number of stomata observed in a high-power field of view.

11. Calculate the total number of stomata on the entire onion leaf by following the directions in Data Table 2.

Data Table 2

Length of leaf portion in mm	= ________ mm
Width of leaf portion in mm	= ________ mm
Calculate area of leaf (length × width)	= ________ mm^2
Calculate number of high-power fields of view on leaf (area of leaf ÷ 0.07 mm^2, the area of one high-power field of view)	= ________
Calculate total number of stomata (number of high-power fields of view × average number of stomata per high-power field of view from Data Table 1)	= ________

Analyze and Conclude

1. Communicating Compare your data with those of your other classmates. Offer several reasons why your total number of stomata for the leaf may not be identical to your classmates'.

____

2. Thinking Critically Analyze the following steps of this experiment and explain how you can change the procedure to improve the accuracy of your data.

a. five trials in Data Table 1

____

b. using 0.07 mm^2 as the area of your high-power field of view

____

3. Concluding Would you expect all plants to have the same number of stomata per high-power field of view? Explain your answer.

____

____

4. Comparing and Contrasting What are the advantages to using sampling techniques? What are some limitations?

____

____

MiniLab 24-1 Growing Plants Asexually

Experimenting

Plants are capable of reproducing asexually. Reproductive cells such as egg or sperm are not needed in asexual reproduction. Plants are able to use structures such as roots, stems, and even leaves to produce new offspring.

Procedure

1. Prepare three different plant parts for study using diagrams A, B, and C on page 654 of your text as a guide.
2. Observe any changes that occur to your plants over the next two weeks.
3. Design a data table that will provide enough room for diagrams of your observations. The number of days since the start of the experiment should be included.
4. Make your initial diagrams of the plant parts today and label these diagrams as "Day 1."
5. Observations should be made every three days. Replace any lost water as needed.

Analysis

1. What experimental evidence do you have that:

a. plants use a variety of structures for asexual reproduction?

____

____

b. asexual reproduction is a rapid process?

____

c. asexual reproduction requires only one parent?

____

2. Describe several advantages of asexual reproduction in plants.

____

____

____

____

Name Date Class

MiniLab 24-2

Observing

Looking at Germinating Seeds

Seeds are made up of a plant embryo, a seed coat, and in some plants, a food-storage tissue. Monocot and dicot seeds differ in their internal structures.

Procedure

1. Obtain from your teacher a soaked, ungerminated corn kernel (monocot), a bean seed (dicot), and corn and bean seeds that have begun to germinate.
2. Remove the seed coats from each of the ungerminated seeds, and examine the structures inside. Use low-power magnification. Locate and identify each structure of the embryo and any other structures you observe.
3. Examine the germinating seeds. Locate and identify the structures you observed in the dormant seeds.

Analysis

1. Diagram the dormant embryos in the soaked seeds, and label their structures.

2. Diagram the germinating seeds, and label their structures.

3. List at least three major differences you observed in the internal structures of the corn and bean seeds.

___

___

___

Examining the Structure of a Flower

Chapter 24

Preparation

Problem

What do the parts of a flower look like? How are they arranged?

Objectives

In this BioLab, you will:

- **Observe** the structures of a flower.
- **Identify** the functions of flower parts.

Materials

flower—any complete flower that is available locally, such as phlox, lily, or tobacco flower
hand lens (or stereomicroscope)
colored pencils (red, green, blue)
2 microscope slides
water
dropper
2 coverslips
microscope
single-edged razor blade

Safety Precautions

Always wear goggles in the lab. Handle the razor blade with extreme caution. Always cut away from you. Use caution when working with a microscope and slides. Wash your hands with soap and water after handling plant material.

Skill Handbook

Use the **Skill Handbook** if you need additional help with this lab.

Procedure

1. Examine your flower. Locate the sepals and petals. Note their numbers, size, color, and arrangement on the flower stem.
2. Remove the sepals and petals from your flower by gently pulling them off the stem. Locate the stamens, each of which consists of a thin filament with a pollen-filled anther on the tip. Note the number of stamens.
3. Locate the pistil. The stigma at the top of the pistil is often sticky. The style is a long, narrow structure that leads from the stigma to the ovary.
4. Place an anther from one of the stamens onto a microscope slide and add a drop of water. Cut the anther into several pieces with the razor blade. **CAUTION:** ***Always take care when using a razor blade.***
5. Examine the anther under low and high power of your microscope. The small, dotlike structures are pollen grains.
6. Slice the ovary in half lengthwise with the razor blade. Mount one half, cut side facing up, on a microscope slide.
7. Examine the ovary section with a hand lens or stereomicroscope. The many small, dotlike structures that fill the two ovary halves are ovules. Each ovule contains an egg cell that is not visible under low power. A tiny stalk connects each ovule to the ovary wall.
8. Identify the ovary and ovules.
9. Make a diagram of the flower, labeling all its parts. Color the female reproductive parts red. Color the male reproductive parts green. Color the remaining parts blue.

Name Date Class

Chapter 24

Examining the Structure of a Flower, *continued*

Analyze and Conclude

1. **Observing** How many stamens are present in your flower? How many pistils, ovaries, sepals, and petals?

2. **Comparing and Contrasting** Make a reasonable estimate of the number of pollen grains in the anther and the number of ovules in an ovary of your flower.

3. **Interpreting Data** Which produces more? Pollen grains by one anther? Ovules produced by one ovary? Give a possible explanation for your answer.

MiniLab 25-1

Observing and Inferring

Observing Animal Characteristics

Animals come in a variety of sizes and shapes, and can be found living in a number of different habitats.

Procedure

1. Use the data table below.
2. Add a few bristles from an old toothbrush to a glass slide. Add a drop of water containing rotifers to your slide. The drop should cover the bristles. Add a coverslip. **CAUTION:** ***Use caution when working with a microscope, slides, and coverslips.***
3. Observe your rotifers under low-power magnification.
4. Use the data table to record the characteristics that you were able to see. Describe the evidence for each trait.

Data Table

Animal Characteristic	Observed? (Yes or No)	Evidence
Multicellular		
Feeding		
Movement		
Size in mm		

Analysis

1. Are these organisms autotrophs or heterotrophs?

2. Were you able to see evidence of feeding? Explain.

3. Are rotifers multicellular? Explain.

Name Date Class

MiniLab 25-2

Observing and Inferring

Check Out a Vinegar Eel

Vinegar eels are roundworms with pseudocoeloms. They exhibit an interesting pattern of locomotion because they have only longitudinal (lengthwise) muscles.

Procedure

1. Prepare a wet mount of vinegar eels. **CAUTION:** ***Use caution when working with a microscope and slides.***
2. Observe them under low-power magnification.
3. Note their pattern of locomotion. Prepare a series of diagrams that illustrate their pattern of movement.
4. Time how fast they move by timing in seconds how long it takes for one roundworm to move across the center of your field of view. Find out the diameter of your low-power field in mm. Calculate vinegar eel speed in mm/sec. You may want to time several animals and average their speed.

Analysis

1. Name the type of symmetry present in vinegar eels.

2. Describe the pattern of locomotion for vinegar eels.

3. How does the pseudocoelom aid vinegar eels in locomotion?

4. What is the speed of locomotion for a vinegar eel? Based on the speed of your vinegar eel, predict the speed in mm/sec for a flatworm. Explain your answer.

Name Date Class

Chapter 25

Zebra Fish Development

PREPARATION

Problem

What do the developmental stages of the zebra fish look like?

Objectives

In this BioLab, you will:

- **Observe** stages of zebra fish development.
- **Record** all observations in a data table.
- **Use the Internet** to collect and compare data from other students.

Skill Handbook

Use the **Skill Handbook** if you need additional help with this lab.

Materials

aquarium
zebra fish
turkey baster
beaker
dropper
petri dish
binocular microscope
wax pencil or labels

Safety Precautions

Always wear safety goggles in the lab. Use caution when working with a binocular microscope and glassware.

PROCEDURE

1. Use the data table
2. Using the turkey baster, draw up water containing zebra fish embryos from the bottom of the aquarium.
3. Release the water into a beaker and allow the embryos to settle to the bottom.
4. Label a petri dish with your name and class period. Fill the bottom of the dish with aquarium water. Using a dropper, place several embryos in your dish.
5. Your teacher will advise you as to the approximate time that fertilization took place. All ages should be reported in your data table as hpf (hours past fertilization).
6. Using a binocular microscope, observe the embryos. Make a diagram in your data table of the embryos' appearances and indicate the age of the embryos in hpf.
7. Go to the Glencoe Science Web Site at **www.glencoe.com/sec/science** to **post your data.**
8. Continue to observe your embryos daily for a minimum of one week. Note the appearance of new organs and when movement is first seen. If you wish to continue watching developmental changes, consult with your teacher for directions. **CAUTION:** ***Wash your hands with soap and water immediately after completing observations.***

Zebra Fish Development, *continued*

Data Table

Date	hpf	Diagram	Observations

Analyze and Conclude

1. **Communicating** Explain why zebra fish are ideal animals for studying embryonic development.

2. **Thinking Critically** Explain why you may not have been able to see stages such as a blastula or gastrula.

3. **Thinking Critically** Suggest how you could change the experiment's design to allow for observing these stages.

4. **Using the Internet** Visit the Glencoe Science Web Site for links to Internet sites that will help you complete sequences of the major changes during development of zebra fish.
   a. between 1 and 10 hpf. Include labeled diagrams of these changes.
   b. between 10 and 28 hpf. Include labeled diagrams.
   c. between 28 and 72 hpf. Include labeled diagrams.

Name Date Class

MiniLab 26-1 Watching Hydra Feed

Observing

Hydras are freshwater cnidarians. They show the typical polyp body plan and symmetry associated with all members of this phylum. Observe how they capture their food.

Procedure

1. Use a dropper to place a hydra into a watch glass filled with water. Wait several minutes for the animal to adapt to its new surroundings. **CAUTION:** ***Use caution when working with a microscope and glassware.***
2. Observe the hydra under low-power magnification.
3. Formulate a hypothesis as to how this animal obtains its food and/or catches its prey.
4. Place brine shrimp in a culture dish of freshwater to avoid introducing salt into the watch glass.
5. Add a drop of brine shrimp to the watch glass while continuing to observe the hydra through the microscope.
6. Note which structures the hydra uses to capture food.

Analysis

1. Describe how the hydra captures food.

____

____

2. Was your hypothesis supported or rejected?

____

3. Sequence the events that take place when a hydra captures and feeds upon its prey.

____

____

4. Explain how your observations support the fact that hydras have both nervous and muscular systems.

____

____

Name Date Class

MiniLab 26-2

Observing

Observing the Larval Stage of a Pork Worm

You can observe the larval stage of a pork worm (*Trichinella spiralis*) embedded within the muscle tissue of its host. It will look like a curled up hot dog surrounded by muscle tissue.

Procedure

1. Examine a prepared slide of pork worm larvae under the low-power magnification of your microscope.
2. Locate several larvae by looking for "spiral worms enclosed in a sac." All other tissue is muscle.
3. Estimate the size of the larva in µm.
4. Diagram one larva. Indicate its size on the diagram.

Analysis

1. Describe the appearance of a pork worm larva.

____

____

2. Why might it be difficult to find larva embedded in muscle when meat inspectors use visual checking methods in packing houses to screen for pork worm contamination?

____

____

____

3. Suggest what inspectors might do to help detect pork worm larvae.

____

____

____

Chapter 26

Observing Planarian Regeneration

PREPARATION

Problem

How can you determine if the flatworm *Dugesia* is capable of regeneration?

Objectives

In this BioLab, you will:

- **Observe** the flatworm, *Dugesia*.
- **Conduct** an experiment to determine if planarians are capable of regeneration.

Materials

planarians
petri dish
springwater
camel hair brush
chilled glass slide
binocular microscope
marking pencil or labels
single-edged razor blade

Safety Precautions

Always wear goggles in the lab. Use extreme caution when cutting with a razor blade. Wash your hands both before and after working with planarians.

Skill Handbook

Use the **Skill Handbook** if you need additional help with this lab.

PROCEDURE

1. Obtain a planarian and place it in a petri dish containing a small amount of springwater. You can pick up a planarian easily with a small camel hair brush.
2. Use a binocular microscope to observe the planarian. Locate the animal's head and tail region and its "eyes." Use diagram **A** on page 735 of your text as a guide
3. Place the animal on a chilled glass slide. This will cause it to stretch out.
4. Place the slide onto the microscope stage. While observing the worm through the microscope, use a single-edged razor to cut the animal in half across the midsection. Use diagram **B** on page 735 of your text as a guide.
5. Remove the head end and place it in a petri dish filled with springwater. Label the dish with the date, your name, and the word "head."
6. Add the tail section to a different petri dish and label it as in step 5, marking this dish "tail."
7. Repeat steps 3–6 with a second flatworm and add the correct pieces to the proper petri dishes.
8. Place the petri dishes in an area designated by your teacher.
9. Prepare a data table that will allow you to record the appearance of your flatworms every other day for two weeks. Include diagrams and the number of days since starting the experiment in your data table.
10. Observe your animals under a binocular microscope and record observations and diagrams in your data table.

Name Date Class

Observing Planarian Regeneration, *continued*

Chapter 26

Analyze and Conclude

1. **Knowledge** To what phylum do flatworms belong? Are planarians free-living or parasitic? What is your evidence?

2. **Observing** What new part did each original head piece regenerate? What new part did each original tail piece regenerate?

3. **Observing** Which section, head or tail, regenerated new parts faster?

4. **Interpreting** Are planarians able to regenerate new parts? Would regeneration be by mitosis or meiosis? Explain.

5. **Thinking Critically** What might be the advantage for an animal that can grow new body parts through regeneration?

6. **Thinking Critically** Would the term "clone" be suitable in reference to the newly formed planarians? Explain your answer.

Name Date Class

MiniLab 27-1

Comparing and Contrasting

Identifying Mollusks

Have you ever taken a walk on the beach and filled your pockets with shells, and as you examined them later, wondered what they were? Use the following dichotomous key to determine the names of the shells.

Procedure

Refer to p. 746 of the text to examine the pictured shells.

1 To use a dichotomous key, begin with a choice from the first pair of descriptions.

2 Follow the instructions for the next choice. Notice that either a scientific name can be found at the end of each description, or directions will tell you to go on to another numbered set of choices.

1A One shell .... Gastropods see 2
1B Two shells .... Bivalves see 3
2A Flat coil .... Sundial shell: *Architectonica nobilis*
2B Thick coil .... see 4
3A Shelf inside shell .... Common Atlantic slipper: *Crepidula fornicata*
3B No shelf inside shell .... see 5
4A Spotted surface .... Junonia shell: *Scaphella junonia*
4B Lined surface .... Banded tulip shell: *Fasciolaria hunteria*
5A Polished surface .... Sunray shell: *Macrocallista mimbosa*
5B Rough surface .... Lion's paw shell: *Lyropecten nodosus*

Analysis

1. Why is a dichotomous key used for a variety of organisms?

2. What shell features were easy to pick out using the key? What features were more difficult?

3. What general feature was used to identify shells?

MiniLab 27-2

Interpreting Scientific Diagrams

A Different View of an Earthworm

What does an earthworm look like internally? You could look at it many different ways—from the dorsal or ventral side, along the length of the animal (a longitudinal view), or in cross section through a segment.

Procedure

1. Diagram A on page 750 of your text illustrates a longitudinal dorsal view of the internal organs of an earthworm. Note that the segments are numbered.
2. Use Diagram B on page 750 of your text as a guide to how a cross-section slice appears through segment 9.

Analysis

Make your own cross-section diagrams of segments 8 and 12. Label all the parts shown in your diagrams.

Name Date Class

How do earthworms respond to their environment?

Chapter 27

Preparation

Problem

How do earthworms respond to light, different surfaces, moist and dry environments, and warm and cold environments?

Hypotheses

Place your worm in a tray with some moist soil. Watch your worm for about 5 minutes, and record what you observe. Make a hypothesis based on your observations about what the worm might do under conditions of light and dark, rough and smooth surfaces, moist and dry surfaces, and warm and cold conditions. Limit your investigation as time requires.

Objectives

In this BioLab, you will:

- **Measure** the sensitivity of earthworms to different stimuli, including light, water, and temperature.
- **Interpret** earthworm responses according to terms of adaptations that promote their survival.

Possible Materials

live earthworms
glass pan
culture dishes
thermometer
dropper
ice
black paper
hand lens or stereomicroscope
paper towels
sandpaper
warm tap water
water
penlight
ruler
cotton swabs

Safety Precautions

Be sure to treat the earthworm in a humane manner at all times. Wet your hands before handling earthworms. Always wear goggles in the lab.

Skill Handbook

Use the **Skill Handbook** if you need additional help with this lab.

Plan the Experiment

1. As a group, make a list of possible ways you might test your hypothesis. Keep the available materials in mind as you plan your procedure.
2. Be sure to design an experiment that will test one variable at a time. Plan to collect quantitative data. Make sure to incorporate a control.
3. Record your procedure and list materials and amounts you will need. Design and construct a data table for recording your findings.

Check the Plan

Discuss the following points with other group members.

1. What data will you collect, and how will they be recorded?
2. Does each test have one variable and a control? What are they?
3. Each test should include measurements of some kind. What are you measuring in each test?
4. How many trials will you run for each test?
5. Assign roles for this investigation.
6. ***Make sure your teacher has approved your experimental plan before you proceed further.***
7. Carry out your experiment. **CAUTION:** ***Return earthworms to the container the teacher has provided.***

Name Date Class

DESIGN YOUR OWN BioLab

How do earthworms respond to their environment?, *continued*

Chapter 27

ANALYZE AND CONCLUDE

1. **Checking Your Hypothesis** Which surface did the worm prefer? Explain.

2. **Interpreting Observations** In which temperature was the worm most active? Explain.

3. **Observing and Inferring** How did the earthworm respond to light? Of what survival value is this behavior?

4. **Observing and Inferring** How did the earthworm respond to dry and moist environments? Of what survival value is this behavior?

5. **Drawing Conclusions** Were your hypotheses supported by your data? Why or why not?

Name Date Class

MiniLab 28-1

Comparing and Contrasting

Crayfish Characteristics

There are more species of arthropods than all of the other animal species combined. This phylum includes a variety of adaptations that are not found in other animal phyla.

Procedure

1. Examine a preserved crayfish. **CAUTION:** ***Wear disposable latex gloves and use a forceps when handling preserved material.***
2. Prepare a data table with the following arthropod traits listed: body segmentation, jointed appendages, exoskeleton, sense organs, jaws.
3. Observe the crayfish. Fill in your data table, indicating which of the arthropod traits you observed.
4. Gently lift the edge of the body covering where the legs attach to the body. Look for feathery structures. These are gills and are part of the animal's respiratory system. **CAUTION:** ***Wash hands with soap and water after handling preserved materials.***

Analysis

1. Do crayfish have all of the traits listed above?

____

____

2. Make a hypothesis as to how crayfish locate food.

____

____

Name Date Class

MiniLab 28-2

Comparing and Contrasting

Comparing Patterns of Metamorphosis

Insects undergo a series of developmental changes called metamorphosis. But not all insects follow the same pattern of metamorphosis.

Procedure

1. Use the data table below.
2. Examine the three life stages of a grasshopper. Complete the information called for in your data table. **CAUTION:** ***Wear disposable latex gloves and use forceps to handle preserved insects.***
3. Examine the four life stages of a moth. Complete the information called for in your data table.

Data Table

Insect	Grasshopper			Moth			
Stage	**egg**	**nymph**	**adult**	**egg**	**larva**	**pupa**	**adult**
Locomotion Method							
Feeding Method							
Able to Reproduce							

Analysis

1. What are the differences between the stages of metamorphosis of a grasshopper and those of a moth?

____

____

2. Correlate the ability to move with ability to feed.

____

____

3. Compare a nymph stage with an adult stage.

____

____

DESIGN YOUR OWN BioLab

Chapter 28

Will salt concentration affect brine shrimp hatching?

PREPARATION

Problem

How can you determine the optimum salt concentration for the hatching of brine shrimp eggs?

Hypothesis

Decide on one hypothesis that you will test. Your hypothesis might be that increased salt concentrations result in an increase in the number of eggs hatched.

Objectives

In this BioLab, you will:

- **Analyze** how salt concentration may affect brine shrimp hatching.
- **Interpret** your experimental findings.

Possible Materials

beaker or plastic bottles
labels or marking pencil
graduated cylinder
brine shrimp eggs
clear plastic trays
salt (noniodized)
balance
water

Safety Precautions

Wear protective eye goggles when preparing solutions.

Skill Handbook

Use the **Skill Handbook** if you need additional help with this lab.

PLAN THE EXPERIMENT

1. Decide on a way to test your group's hypothesis. Keep the available materials in mind as you plan your procedure. Be sure to include a control. For example, you might place brine shrimp eggs in two trays—one with the salt concentration of the water brine shrimp normally inhabit, and one with a different salt concentration.
2. Decide how long you will make observations and how you will judge the extent of egg hatching.
3. Decide on the number of different salt water concentrations to use and what these concentrations will be. Review the steps needed to prepare solutions of different concentrations.

Check the Plan

Discuss the following points with other group members to decide on the final procedure for your experiment.

1. What is your one independent variable? Your dependent variable?
2. What will be your control?
3. How much water will you add to each tray and how will you measure the same number of eggs to be used in each tray?
4. Will it be necessary to control variables such as light and temperature?
5. What data will you collect and how will it be recorded?
6. ***Make sure your teacher has approved your experimental plan before you proceed further.***
7. Carry out your experiment.

Name Date Class

DESIGN YOUR OWN BioLab

Will salt concentration affect brine shrimp hatching?, *continued*

Chapter 28

ANALYZE AND CONCLUDE

1. **Interpreting Data** Using specific numbers from your data, explain how salt concentration affects brine shrimp hatching.

2. **Drawing a Conclusion** Was your hypothesis supported? Explain.

3. **Identifying and Controlling Variables** What were the independent and dependent variables? What were some of the variables that had to be controlled?

4. **Hypothesizing** Formulate a hypothesis that explains why high salt concentrations may be harmful to brine shrimp hatching.

5. **Classifying** Classify brine shrimp. Identify their kingdom, phylum, class, order, family, genus, and species.

Name Date Class

MiniLab 29-1

Observing and Inferring

Examining Pedicellariae

Echinoderms move by tube feet. They also have tiny pincers on their skin called pedicellariae.

Procedure

1. Observe a slide of sea star pedicellariae under low-power magnification. **CAUTION:** ***Use caution when working with a microscope and slides.***
2. Record the general appearance of one pedicellaria. What does it look like?
3. Make a diagram of one pedicellaria under low-power magnification.
4. Indicate the size of one pedicellaria in micrometers.

Analysis

1. Describe the general appearance of one pedicellaria.

____

____

2. What is the function of this structure?

____

____

3. Explain how the structure of pedicellariae assists in their function.

____

____

____

____

____

MiniLab 29-2

Observing

Examining a Lancelet

Branchiostoma californiense is a small, sea-dwelling lancelet. At first glance, it appears to be a fish. However, its structural parts and appearance are quite different.

Procedure

1. Place the lancelet onto a glass slide. **CAUTION:** ***Wear disposable latex gloves and handle preserved material with forceps.***
2. Use a dissecting microscope to examine the animal. **CAUTION:** ***Use care when working with a microscope and slides.***
3. Prepare a data table that will allow you to record the following: General body shape, Length in mm, Head region present, Fins and tail present, Nature of body covering, Sense organs such as eyes present, Habitat, Segmented body.
4. Indicate on your data table if the following can easily be observed: gill slits, notochord, dorsal hollow nerve cord.

Analysis

1. How does *Branchiostoma* differ structurally from a fish? How are its general appearance and habitat similar to those of a fish?

2. Explain why you were not able to see gills, notochord, and a dorsal hollow nerve cord.

3. Using its scientific name as a guide, where might the habitat of this species be located?

Name Date Class

Observing Sea Urchin Gametes and Egg Development

Chapter 29

PREPARATION

Problem

How can you induce a sea urchin to release its gametes?

Objectives

In this BioLab, you will:

- **Induce** sea urchins to release their gamete cells.
- **Observe** living sperm and egg cells under the microscope.
- **Observe** developmental changes in a fertilized sea urchin egg.

Materials

live sea urchins
sea water
glass slides and coverslips
syringe filled with potassium chloride
beakers
petri dish
dropper
microscope
test tube

Safety Precautions

Always wear goggles in the lab.

Skill Handbook

Use the **Skill Handbook** if you need additional help with this lab.

PROCEDURE

1. Fill a small beaker (250mL) with sea water.
2. Obtain a live sea urchin from your teacher and locate an area of soft tissue next to its mouth.
3. Using a syringe, your teacher will insert the needle into this soft tissue and inject the syringe contents into the sea urchin.
4. Turn your animal so that its mouth is facing up and place it in a petri dish. **CAUTION:** ***Use care in handling live animals.***
5. Wait a minute or two, then check the petri dish. If the sea urchin is male, a milky white mass of sperm will be present in the dish. If it is female, a yellow orange mass of eggs will be seen.
6. If you have a female sea urchin, hold her upside down directly over the seawater-filled beaker and allow the eggs to fall directly into the water.
7. If your urchin is male, use a dropper to add several drops of sperm from the petri dish to your beaker of sea water.
8. Check with your classmates to see who has a male and who has a female sea urchin. Share gamete cells.
9. Use a clean dropper to transfer a drop of sperm from the beaker to a microscope slide. Observe under low power without a coverslip.
10. Add a coverslip and observe under high power. Note the movement of sperm. Draw several sperm cells and indicate their size in µm. Note the approximate number of sperm cells present.
11. Repeat steps 9 and 10 for egg cells. In step 10, use only low power to observe egg cells.
12. For this step, work with a partner. While one partner transfers some sperm to the slide with egg cells using a clean dropper, the other partner should observe under low power.
13. Observe the process of fertilization and note any changes that occur to the egg. Record your observations in a data table.

Name Date Class

Observing Sea Urchin Gametes and Egg Development, *continued*

Chapter 29

14. When fertilization has been accomplished, place the fertilized eggs in a test tube filled with 10 mL of seawater. Label your tube and observe the eggs 24 hours later under low power. Record any changes that you see. **CAUTION:** ***Wash your hands immediately after working with animals.***

Analyze and Conclude

1. **Compare and Contrast** Compare eggs and sperm, noting numbers released, numbers observed under low power, size, and ability to move.

____

____

____

2. **Predicting** Based on the pattern of fertilization, predict the reason for the large number of gametes released in nature.

____

____

____

3. **Observing** Describe the behavior of sperm when they first come in contact with an egg.

____

____

____

4. **Observing** How does an unfertilized egg differ in appearance from a fertilized egg? Draw both eggs in your data table.

____

____

____

MiniLab 30-1

Experimenting

Measuring Breathing Rate in Fishes

Fishes are able to extract oxygen from water as it flows over their gills. Their rate of breathing is related to the availability of oxygen in the water. More oxygen results in a slower breathing rate. The breathing rate of a fish can be estimated by counting the number of times per minute its gill covers open to allow water to flow across its gills.

Procedure

1. Make a hypothesis about how the breathing rate of a fish may be influenced by a change in temperature of the water in which it is swimming. Record your hypothesis.
2. Fill a beaker with non-chlorinated water. Let the beaker sit until the water reaches room temperature (about 20°C). Measure the temperature with a thermometer and record it in a data table.
3. Add a small goldfish. **CAUTION:** ***Handle animals with care.*** Wait 5 minutes for the fish to acclimatize to the water temperature.
4. Count the number of times the goldfish's gill covers open in one minute. This is a measure of the rate at which the fish is breathing. Record your result in a data table.
5. Repeat step 4 four more times. Find an average for your trials and record these data in your table.
6. Remove the goldfish from the beaker. Add one ice cube made from non-chlorinated water to the beaker.
7. When the ice cube has melted, record the water's temperature.
8. Repeat steps 3–5. Remove goldfish from the beaker, and add another ice cube to the beaker. Repeat step 7.
9. Repeat steps 3–5. **CAUTION:** ***Wash your hands after working with animals.***

Analysis

1. How does water temperature influence the rate at which breathing occurs in a goldfish?

2. Do your data support your hypothesis? Explain.

MiniLab 30-1

Measuring Breathing Rate in Fishes, *continued*

3. Was the breathing rate faster or slower in colder water compared with warmer water?

____

4. How might the amount of oxygen in cold water compare with that in warm water? Explain your answer.

____

____

____

____

5. Sequence the events associated with a fish obtaining oxygen. Start with a molecule of oxygen in water and be sure to include the capillaries located in the fish's gills.

____

____

____

____

Name Date Class

MiniLab 30-2

Comparing and Contrasting

Looking at Frog and Tadpole Adaptations

An adult frog and its larval stage are adapted to different habitats. How are the structures of a frog and a tadpole adapted to their environments?

Procedure

1. Use the data table below.
2. Examine a living or preserved adult frog and larval (tadpole) stage. **CAUTION:** ***Wear disposable latex gloves and use a forceps when handling preserved specimens.***
3. Observe the first seven traits listed. Complete your data table for these observations.
4. Use references to fill in the information for the last three traits listed.

Data Table

Trait of information	Tadpole	Adult
Limbs present?		
Eyes present?		
Tympanic membrane present?		
Tail present?		
Mouth present?		
Nature of skin (color and texture)		
General size		
Respiratory organ type		
Diet		
Habitat		

Analysis

1. Explain how hind leg musculature aids in adult frog survival.

2. Correlate the type of respiratory organ in an adult and a tadpole with their differing habitats.

MiniLab 30-2

Looking at Frog and Tadpole Adaptations, *continued*

3. Correlate the type of appendages (arm, leg, tail) in an adult and a tadpole with their differing habitats.

____

____

4. Correlate mouth size in an adult and a tadpole with their differing diet.

____

____

5. Explain how eyes may aid in the survival of both stages.

____

____

6. Explain why the tympanic membrane may not be essential to the survival of a tadpole.

____

____

7. Predict how skin color and texture aids in adult frog survival. **CAUTION:** ***Wash your hands after working with live or preserved animals.***

____

____

Chapter 30

Development of Frog Eggs

PREPARATION

Problem

How does temperature affect the development of frog eggs?

Objectives

In this BioLab, you will:

- **Compare** development of frog eggs at varying temperatures.
- **Distinguish** among various stages of development.

Materials

Ringer's solution
4 culture dishes
lightbulbs with source of electricity
thermometer
binocular microscope
frog eggs, *Xenopus laevis*
flashlight

Safety Precautions

Always wear goggles in the lab. Wash hands before and after each observation.

Skill Handbook

Use the **Skill Handbook** if you need additional help with this lab.

PROCEDURE

1. Obtain four culture dishes of Ringer's solution and fertilized frog eggs.
2. Make a data table similar to the one shown for sketching of stages of development of the eggs.
3. Observe your eggs and determine their stage of development. At room temperature, you should see the two-cell stage about 1.5 hours after fertilization, the eight-cell stage at 2.25 hours, the 32-cell stage at 3 hours, the late gastrula stage at 9 hours, and a visible head area between 18 and 20 hours. Hatching will occur at about 50 hours (about two days).
4. Set up the appropriate numbers of lightbulbs of different wattages over the water to keep the temperatures at 20°, 25°, and 30°C.
5. Place a dish of eggs in the refrigerator. Measure the temperature of your refrigerator. Keep a flashlight on in the refrigerator at all times so that you have only one variable, the temperature.
6. Make a hypothesis about how temperature will affect development of the eggs.
7. Set up a time schedule for observations and making sketches based on what stage you are observing. Make observations until the eggs hatch. Record your observations in a journal.
8. Observe your eggs under the microscope according to the schedule you have made. Draw sketches of your eggs in the data table.

Data Table

Temperature	Day 1	Day 2	Day 3
30° C			
25° C			
20° C			
refrigerator			

INVESTIGATE BioLab

Chapter 30

Development of Frog Eggs, *continued*

Analyze and Conclude

1. **Interpreting Observations** Which eggs develop the fastest? The slowest? Explain.

2. **Interpreting Observations** Did your data support your hypothesis? Explain.

3. **Drawing Conclusions** What advantage is it for frogs to have eggs that develop at different rates that correspond to different temperatures?

4. **Thinking Critically** What would happen if frog eggs developed when the weather was still cold in the spring?

MiniLab 31-1

Comparing and Contrasting

Comparing Feathers

Birds have two kinds of feathers. Contour feathers used for flight are found on a bird's body, wings, and tail. Down feathers lie under the contour feathers and insulate the body.

Procedure

1. Examine a contour feather with a hand lens, and make a sketch of how the feather filaments are hooked together.
2. Examine a down feather with a hand lens. Draw a diagram of the filaments of the down feather.
3. Fan your face with each feather separately. Note how much air is moved past your face by each type of feather. **CAUTION:** ***Wash your hands after handling animal material.***

Analysis

1. How does the structure of a contour feather help a bird fly?

____

____

2. How does the structure of a down feather keep a bird warm?

____

____

____

____

3. How can you explain the differences you felt when fanning with each feather?

____

____

____

____

Name Date Class

MiniLab 31-2

Comparing and Contrasting

Feeding the Birds

In the winter, it may be difficult for some birds to find food, especially if you live in an environment often blanketed with snow. Making a bird feeder and watching birds feed can be an enjoyable activity for you that may save some birds from starvation. If you do begin feeding birds in the winter, continue to feed them until natural food again becomes available in the spring.

Procedure

1. Obtain several large, plastic milk bottles. Cut two holes 5 cm from the base on opposite sides of each bottle, each about 8 cm^2. These are the openings birds will use to find the food inside.
2. Place small drainage holes in the bottom of each bottle. Hang the bottles from wires strung through small holes in the neck of each one.
3. Place a different kind of seed (sunflower seeds, hulled oats, cracked corn, wheat, thistle, millet) in different bottles. Add new seed when needed.
4. Using a bird guide, make a list of numbers and kinds of birds that frequent each feeder, noting the type of food offered.

Analysis

1. What type of seed attracted the largest variety of bird types?

___

2. Did any birds visit more than one feeder?

___

___

3. What do you think an ideal bird food would be?

___

___

Name Date Class

Chapter 31

Which egg shape is best?

PREPARATION

Problem

What shape would be best for an egg to reduce the distance it could roll if pushed from a nest?

Hypotheses

There are several hypotheses that you can test. Your hypothesis might be that egg shape influences the distance an egg rolls, or that shape determines the tightness of circular rolling patterns.

Objectives

In this Biolab, you will:

- **Design** an experiment to test your hypotheses.
- **Model** different egg shapes and egg masses.
- **Experiment** to test your hypotheses.
- **Draw conclusions** based on your experimental data.

Possible Materials

clay
cardboard ramp
ruler
string
hard-boiled egg
Ping-Pong ball
golf ball
balance
protractor

Safety Precautions

Always wear goggles in the lab.

Skill Handbook

Use the **Skill Handbook** if you need additional help with this lab.

PLAN THE EXPERIMENT

1. Decide on a way to test your group's hypothesis. Keep the list of available materials in mind as you plan your procedure.
2. There are a number of questions to be asked before starting. Here are some suggestions. How will you incorporate a control? How many egg shapes will you test? How will you model your egg shapes? How many trials will you perform? How might you keep egg models identical in mass? How will you measure the angle of the cardboard ramp? Where will you start to measure distance rolled?

Check the Plan

Discuss the following points with other group members to decide the final procedure for each of your experiments.

1. What is your independent and dependent variable?
2. How will you eliminate all other variables?
3. What data will you collect? How many trials will you run?
4. Will you need a data table and how might it be organized?
5. ***Make sure your teacher has approved your experimental plan before you proceed further.***
6. Carry out your experiments.

Name Date Class

DESIGN YOUR OWN BioLab

Which egg shape is best?, *continued*

Chapter 31

Analyze and Conclude

1. **Hypothesizing** Record your hypothesis.

2. **Interpreting Data** Describe your results after testing your hypothesis.

3. **Concluding** Do your data support your hypothesis? Explain using both quantitative and qualitative observations.

4. **Identifying Variables** What were your independent and dependent variables?

5. **Concluding** In general, how does mass influence the distance an egg will roll? How does egg shape influence the distance an egg will roll or the pattern taken when it rolls?

6. **Predicting** Predict why egg shape or mass may be helpful adaptations when considering the variety of habitats where birds live.

MiniLab 32-1

Observing

Anatomy of a Tooth

Most mammals have teeth. Teeth typically have evolved into a variety of shapes, depending on the diet of the animal.

Procedure

1. Examine a prepared slide of a human tooth under low-power magnification. The slide is a longitudinal section view of the tooth. **CAUTION:** ***Use caution when working with a microscope and slides.***
2. Locate the different areas that form the human tooth. From outside to inside they are: enamel, dentine, and pulp cavity. Each area varies in thickness and texture. You will have to move the slide from left to right or up and down to see all areas.
3. Diagram the appearance of the entire cross section as it appears under low power, labeling the three areas.

Analysis

1. Which area is the thickest? Thinnest?

____

2. Describe the nature of the three areas. Use such terms as compact, loose, bonelike, vascular (with blood vessels).

____

____

3. Which area of the three would you expect to be the toughest? Explain why.

____

____

4. Draw cross sections of the two tooth shapes shown on page 869 of your text, at the points marked x and y on each diagram. Include the three areas just studied and label each area.

____

____

Name Date Class

Observing

MiniLab 32-2 Mammal Skeletons

Owls are predators. They feed on small mammals as well as on birds. After eating a meal, the tough indigestible parts of prey, such as bones, are regurgitated as pellets. The skeletons of a variety of small mammals can be studied by examining owl pellets.

Procedure

1. Place an owl pellet onto a sheet of paper toweling.
2. Use a forceps to remove a very small amount of the outer covering.
3. Prepare a wet mount of this material and observe under low-, then high-power magnification. Diagram what you see. Use forceps to open the pellet and remove all bones that are present. Look especially for skulls. (Note: Skulls may be small and certain parts, such as lower jaws, will be separated.)
4. Identify each mammalian skull using Diagrams A and B on page 871 of your text as a guide.
5. Attempt to reconstruct the skeleton for an entire animal. You may wish to glue the skeletal piece onto a piece of cardboard. **CAUTION:** ***Wash your hands after handling animal materials.***

Analysis

1. What was the outer covering on the pellet? What does this tell you about the contents of the pellet? Explain.

____

____

2. How many vole and shrew skulls were present in your pellet?

____

____

3. How were you able to differentiate the skulls of these two mammals?

____

____

4. Predict if voles and shrews are herbivores or carnivores based on appearance of their teeth. Explain.

____

____

Name Date Class

Chapter 32

Domestic Dogs Wanted

PREPARATION

Problem
What are the most popular breeds of dog?

Objectives
In this BioLab, you will:
- **Observe** the characteristics of dogs.
- **Record** the popularity of different dog breeds.
- **Use the Internet** to collect and compare data from other students.
- **Compare** the most popular dog breeds with the most common breeds found in animal shelters.

Materials
computer access to the Internet

Skill Handbook
Use the **Skill Handbook** if you need additional help with this lab.

PROCEDURE

1. Use both data tables.
2. Go to the Glencoe Science Web Site at www.glencoe.com/sec/science to find links to sites that rank the most popular breeds of dog. Also find links to animal shelters and pet adoption agencies.
3. From your data, determine the five most popular dog breeds. Record your findings in Data Table I.
4. Find pictures of these dogs and record the physical characteristics unique to their breeds in Data Table I.
5. At the Glencoe Science Web Site find links to five web sites of animal shelters and pet adoption agencies across the country. Many of these sites post pictures and other information about the dogs that are up for adoption.
6. From your data, determine which breeds are most commonly found in animal shelters. Record the five most common breeds in Data Table II.
7. Find pictures of these dog breeds and record their unique characteristics in Data Table II.

Data Table I

Most Popular Dog Breeds		
Rank	**Breed**	**General Characteristics**

Domestic Dogs Wanted, *continued*

Data Table II

Most Common Dog Breeds Found in Animal Shelters		
Rank	**Breed**	**General Characteristics**

Analyze and Conclude

1. **Using the Internet** Compare the data in Tables I and II. Are there similarities? How might you explain any similarities or differences? Consider the time of year, the age of the dog, and any breeds of dog that are popularized by television or other advertising while formulating your answer.

2. **Thinking Critically** Are there any breeds that are popular as pets that are commonly found in animal shelters? Propose an explanation for this.

3. **Comparing and Contrasting** What characteristics do domestic dogs have that are the same as all other carnivores? As all other mammals?

4. **Problem Solving** Propose a local program to reduce the number of stray and abandoned dogs.

Name Date Class

MiniLab 33-1

Experimenting

Testing an Isopod's Response to Light

Isopods, the pill bugs and sow bugs, are common arthropods on sidewalks or patios. They are actually land crustaceans and respire through gill-like organs that must be kept moist at all times.

Procedure

1. Use the data table on the right.
2. Prepare a plastic dish using the diagram on page 890 of your text as a guide. Moisten the paper toweling.
3. Place six isopods in the center of the dish and add the cover. Place the dish near a lamp or next to a classroom window with light. Have the light strike the dish as shown in the diagram. **CAUTION:** ***Treat isopods gently.***
4. Wait five minutes and observe the dish. Count and record in your data table the number of isopods on the dark or light side. This is your "five minute observation."
5. Repeat step 4 three more times, waiting five minutes before each observation.

Data Table

Observation in minutes	Number of isopods present	
	Light side	Dark side
5		
10		
15		
20		
25		

Analysis

1. Do isopods tend to move toward light or dark areas? Support your answer with specific numbers from your data.

2. Might the behavior of isopods toward light or darkness be innate or learned? Explain your answer.

3. What might be the adaptive advantage for the observed isopod behavior and their response to light? Explain how natural selection may have influenced this isopod behavior.

4. Prepare a bar graph that depicts your data.

Name Date Class

MiniLab 33-2

Experimenting

Solving a Puzzle

You are given a bunch of keys and asked to open a door. How do you go about finding the right key? Several attempts are needed and then finally the door opens. The next time you are asked to perform the same task, can you go directly to the correct key? Chances are, you can. You have learned how to solve this problem.

Procedure

1. Use the data table on the right.
2. Obtain a paper puzzle from your teacher.
3. Time how long it takes you to assemble the puzzle pieces into a perfect square.
4. Record the time it took and call this trial 1.
5. Disassemble the square and mix the pieces.
6. Repeat step 3 for four more trials.

Data Table

Trial	Time needed to complete square puzzle
1	
2	
3	
4	
5	

Analysis

1. Using your data, explain how the time needed to complete the puzzle changed from trial 1 to trial 5.

2. Was the final completion of the puzzle an example of innate behavior? Explain your answer.

3. Was the final completion of the puzzle an example of learned behavior? Explain your answer.

4. Analyze your behavior when solving the puzzle as to the role that imprinting, trial and error, conditioning, and insight may have played in improving your trial times.

Name Date Class

Chapter 33

Behavior of a Snail

Preparation

Problem

How can you test the behavior of snails to touch stimuli?

Objectives

In this BioLab, you will:

- **Test** the response of snails to touch.
- **Measure** the time needed for habituation to occur after repeated touch stimuli.

Materials

snails
small dish
dropper
scissors
springwater
dissecting microscope
probe constructed from tape, rubber band, and pencil

Safety Precautions

Always wear goggles in the lab. Wash your hands both before and after handling any animals. Use caution when working with live animals.

Skill Handbook

Use the **Skill Handbook** if you need additional help with this lab.

Procedure

1. Use the data table
2. Prepare a stimulator probe by taping a small piece of a cut rubber band to the tip of a pencil.
3. Cover the bottom of a small dish with springwater.
4. Obtain a snail from your teacher and place it in the dish.
5. Use a dissecting microscope to examine and locate its head. Its head has two antennae that it can extend and retract.
6. Place the dish on your desk.
7. Lightly touch the snail's anterior end using the end of the rubber band probe. Note if it responds (yes or no), and record the direction the snail moves. Consider this trial 1.
8. Repeat step 7 for four more trials.
9. Lightly touch the snail's posterior end using the rubber band probe. Record your observations and conduct a total of five trials.

Data Table

Body area	Response to touch		
Trial	Anterior	Posterior	Middle
1			
2			
3			
4			
5			
Habituation studies			
Rate of stimulation	Number of stimulations needed to reach habituation		

INVESTIGATE BioLab

Behavior of a Snail, *continued*

10. Repeat step 9, touching the middle of the snail's body.
11. Test the snail's ability to become habituated from stimulation to its anterior end.
    a. Continue to touch the snail's anterior end with the probe every 10 seconds until habituation occurs. Continue testing for a reasonable length of time if habituation does not occur.
    b. Count and record the number of stimulations needed for habituation.

Analyze and Conclude

1. **Hypothesizing** Are the responses shown by snails to touch learned or innate? Explain your answer.

2. **Observing** Describe the direction that a snail moves when its anterior and posterior ends are stimulated. Does one end appear to be more sensitive than the other? Is the middle sensitive to touch? Is the speed of response slow or rapid?

3. **Hypothesizing** Explain how the behavior of responding to touch may be an adaptation for survival.

4. **Experimenting** Why did you perform several trails for each experiment involving stimulation of the anterior, posterior, and middle of the snail?

5. **Defining Operationally** Define the term habituation.

6. **Concluding** Explain how your data may be used to support the observation that snails are not easily habituated to touch. Use actual data to support your answer.

7. **Predicting** How might this lack of habituation serve as an adaptation for survival?

Name Date Class

MiniLab 34-1

Comparing

Examine Your Fingerprints

Fingerprints play a major role in any police investigation. Because a fingerprint is an individual characteristic, extensive FBI fingerprint files are used for identification in criminal cases.

Procedure

1. Press your thumb lightly on the surface of an ink pad.
2. Roll your thumb from left to right across the corner of an index card, then immediately lift your thumb straight up from the paper.
3. Repeat the steps above for your other four fingers, placing the prints in order across the card.
4. Examine your fingerprints with a magnifying lens, identifying the patterns in each by comparing them with the diagrams on page 925 of your text.
5. Compare your fingerprints with those of your classmates.

Analysis

1. Are the fingerprint patterns on your five fingers all the same?

____

____

2. Do any of your fingerprints show the same patterns as those of a classmate?

____

____

3. Why is a fingerprint a good way to identify a person?

____

____

MiniLab 34-2

Interpreting

Look at Muscle Contraction

Muscle fibers are composed of a number of small functional units called sarcomeres. Sarcomeres, in turn, are composed of protein filaments called actin and myosin. The sliding action of these filaments in relation to each other results in muscle contraction.

Procedure

1. Look at Diagrams A and B on page 937 of your text. Diagram A shows a sarcomere in a relaxed muscle. Diagram B shows a sarcomere in a flexed muscle.
2. Using a centimeter ruler, measure and record the length of a(n): sarcomere in Diagram A, myosin filament in Diagram A, actin filament in Diagram A. Record your data in a table.
3. Repeat step 2 for Diagram B.

Analysis

1. When a muscle contracts, do actin or myosin filaments shorten? Use specific data from your model to support your answer.

____

____

____

2. How does the sarcomere shorten when parts that make it up don't shorten?

____

____

____

Name Date Class

Chapter 34

Does fatigue affect the ability to perform an exercise?

Preparation

Problem

How does fatigue affect the number of repetitions of an exercise you can accomplish? How do different amounts of resistance affect rate of fatigue?

Hypotheses

Hypothesize whether or not muscle fatigue has any effect on the amount of exercise muscles can accomplish. Consider whether fatigue occurs within minutes or hours.

Objectives

In this BioLab, you will:

- **Hypothesize** whether or not muscle fatigue affects the amount of exercise muscles can accomplish.
- **Measure** the amount of exercise done by a group of muscles.
- **Make a graph** to show the amount of exercise done by a group of muscles.

Possible Materials

stopwatch or clock with second hand
graph paper
small weights
wooden box or step stool

Skill Handbook

Use the **Skill Handbook** if you need additional help with this lab.

Plan the Experiment

1. Design a repetitive exercise for a particular group of muscles. Make sure you can count single repetitions of the exercise (for example, one sit-up or one jumping jack) over time.
2. Work in pairs, with one member of the team being a timekeeper and the other member performing the exercise.
3. Consider setting up your experiment so that the amount of resistance is the independent variable. Compare your design with those of other groups.

Check the Plan

1. Be sure the exercises are ones that can be done rapidly and cause a minimum of disruption to other groups in the classroom.
2. Consider how long you will do the activity and how often you will record measurements.

Data Table

Time interval	Number of repetitions
First minute	
Second minute	
Third minute	
Fourth minute	
Fifth minute	

3. ***Make sure your teacher has approved your experimental plan before you proceed further.***
4. Use the data table above to record the number of exercise repetitions per time interval.
5. Carry out the experiment.
6. On a piece of graph paper, plot the number of repetitions on the vertical axis and the time intervals on the horizontal axis.

DESIGN YOUR OWN BioLab

Does fatigue affect the ability to perform an exercise?, *continued*

Chapter 34

ANALYZE AND CONCLUDE

1. **Making Inferences** What effect did repeating the exercise over time have on the muscle group?

2. **Comparing and Contrasting** As you repeated the exercise over time, how did your muscles feel?

3. **Recognizing Cause and Effect** What physiological factors are responsible for fatigue?

4. **Thinking Critically** How well do you think your fatigued muscles would work after 30 minutes of rest? Explain your answer.

Name Date Class

MiniLab 35-1

Interpreting the Data

Evaluate a Bowl of Soup

As a consumer, you are bombarded by advertising that promotes the nutritional benefits of specific food products. Choosing a food to eat on the basis of such ads may not make nutritional sense. By examining the ingredients of processed foods, you can learn important things about their nutritional content.

Table 35.3

Percentage of Daily Value (DV)	
Carbohydrates	60%
Fat	30%
Saturated Fats	10%
Cholesterol	1.5%
Protein	10%
Total Calories	2000

Procedure

1 Examine the information in the table listing the daily value (DV) of various nutrients. DV expresses what percent of Calories should come from certain nutrients. For instance, in the proposed diet of 2000 Calories, 60 percent of the Calories should come from carbohydrates.

2 Examine the nutritional information on the soup can label on page 957 of your text and compare it with the DV table.

Analysis

1. Does your bowl of soup provide more than 30 percent of any of the daily nutrients? Which ones?

____

2. Evaluate the percentage of Calories in soup that are provided by saturated fat.

____

3. Is soup a nutritious meal? Explain your answer.

____

____

Name Date Class

MiniLab 35-2

Observing

Compare Thyroid and Parathyroid Tissue

Although their names seem somewhat similar, the thyroid and parathyroid glands perform rather different functions within the body.

Procedure

1 Use the data table on the right.

2 Use low-power magnification to examine a prepared slide of thyroid and parathyroid endocrine gland tissue. Note: Both tissues appear on the same slide. **CAUTION:** ***Use caution when working with a microscope and prepared slides.***

3 The image on page 964 of your text is a photograph of thyroid and parathyroid tissue. Use it as a guide in locating the two types of endocrine gland tissue under low power and in answering certain analysis questions.

4 Now locate each type of gland tissue under high-power magnification. Draw what you see in the table above. Then use what you learned in the chapter to identify the names of the hormones produced by each gland.

Data Table

Tissue	Drawing	Name of hormone(s) produced
Thyroid		
Parathyroid		

Analysis

1. Compare the microscopic appearance of parathyroid tissue to that of thyroid tissue.

____

____

____

2. a. Which tissue type contains follicles (large liquid storage areas)?

____

b. What may be present within the follicles?

____

c. Are follicles composed of cells? Could they produce the hormone associated with this gland tissue? Explain your answer.

____

d. Hypothesize what the function may be for the thin layer of tissue that surrounds each follicle.

____

3. How might you explain the fact that both thyroid and parathyroid tissue can be seen on the same slide?

____

____

Name Date Class

Average Growth Rate in Humans

Chapter 35

Preparation

Problem

Is average growth rate the same in males and females?

Objectives

In this BioLab, you will:

- **Graph** the average growth rates in males and females.
- **Identify** any differences in the average growth rates of males and females.

Materials

blue pencils
graph paper
red pencils
ruler

Skill Handbook

Use the **Skill Handbook** if you need additional help with this lab.

Procedure

1. Construct a graph for the growth rate data that shows mass on the vertical axis and age on the horizontal axis.
2. On the graph, plot the data shown in the table for the average female growth in mass from ages 8 to 18. Use a ruler to connect the data points with a straight red line.
3. On the same graph, plot the data for the average male growth in mass from ages 8 to 18. Connect these data points with a straight blue line.
4. Construct a second graph that shows height on the vertical axis and age on the horizontal axis.
5. Plot the data for the average female growth in height from ages 8 to 18. Connect the data points with a straight red line.
6. Plot the data for the average male growth in height from ages 8 to 18. Connect these data points with a straight blue line.

Data Table: Averages for growth in humans

	Mass (kg)		Height (cm)	
Age	Female	Male	Female	Male
8	25	25	123	124
9	28	28	129	130
10	31	31	135	135
11	35	37	140	140
12	40	38	147	145
13	47	43	155	152
14	50	50	159	161
15	54	57	160	167
16	57	62	163	172
17	58	65	163	174
18	58	68	163	178

Name Date Class

Average Growth Rate in Humans, *continued*

Chapter 35

Analyze and Conclude

1. **Analyzing Data** During what ages do females and males increase the most in mass? In height?

2. **Analyzing Data** Interpret the data to find if the average growth rate is the same in males and females.

3. **Thinking Critically** How can you explain the differences in growth rates between males and females?

4. **Relating Concepts** Why do you think male and female growth rates increase during the teen years?

Name Date Class

MiniLab 36-1

Experimenting

Distractions and Reaction Time

Have your ever tried to read while someone is talking to you? What effect does such a distracting stimulus have on your reaction time?

Procedure

1. Work with a partner. Sit facing your partner as he or she stands.
2. Have your partner hold the top of a meterstick above your hand. Hold your thumb and index finger about 2.5 cm away from either side of the lower end of the meterstick without touching it.
3. Tell your partner to drop the meterstick straight down between your fingers.
4. Catch the meterstick between your thumb and finger as soon as it begins to fall. Measure how far it falls before you catch it. Practice several times.
5. Run ten trials, recording the number of centimeters the meterstick drops each time. Average the results.
6. Repeat the experiment, this time counting backwards from 100 by fives (100, 95, 90, . . .) as you wait for your partner to release the meterstick.

Analysis

1. Did your reaction time improve with practice? Explain.

____

____

____

2. How was your reaction time affected by the distraction (counting backward)?

____

____

____

3. What other factors, besides distractions, would increase reaction time?

____

____

____

Name Date Class

MiniLab 36-2

Analyzing Information

Interpret a Drug Label

One common misuse of drugs is not following the instructions that accompany them. Over-the-counter medicines can be harmful—even fatal—if they are not used as directed. The Food and Drug Administration requires that certain information about a drug be provided on its label to help the consumer use the medicine properly and safely.

Procedure

1. The photograph on page 991 of your text shows a label from an over-the-counter drug. Read it carefully.
2. Use the data table below. Fill in the table using information on the label.

Information from a drug label

People with these conditions should avoid this drug	Possible side effects	This drug should not be taken with these medicines	Symptoms this drug will relieve	Correct dosage

Analysis

1. What is a side effect? What side effects are caused by this drug?

2. Why should a person never take more than the recommended dosage?

3. How are over-the-counter drugs different from prescription drugs?

Name Date Class

Chapter 36

What drugs affect the heart rate of *Daphnia*?

Preparation

Problem

What legally available drugs are stimulants to the heart? What legal drugs are depressants? Because these drugs are legally available, are they less dangerous?

Hypotheses

Based on what you learned in this chapter, which of the drugs listed under Possible Materials do you think are stimulants? Which are depressants? How will they affect the heart rate in *Daphnia*? Make a hypothesis concerning how each of the drugs listed will affect heart rate.

Objectives

In this BioLab, you will:

- **Measure** the resting heart rate in *Daphnia.*
- **Compare** the resting heart rate with the heart rate when a drug is applied.

Possible Materials

aged tap water
Daphnia culture
dilute solutions of coffee, tea, cola, ethyl alcohol, tobacco, and cough medicine (destromethorphen)
dropper
microscope
microscope slide

Safety Precautions

Do not drink any of the solutions used in this lab. Always wear goggles in the lab. Use caution when working with a microscope, microscope slides, and glassware.

Skill Handbook

Use the **Skill Handbook** if you need additional help with this lab.

Plan the Experiment

1. Using a dropper, place a single *Daphnia* crustacean on a slide.
2. Observe the animal on low power and find its heart.
3. Design an experiment to measure the effect on heart rate of four of the drug-containing substances in the Possible Materials list.
4. Design and construct a data table for recording your data.

Check the Plan

1. Be sure to consider what you will use as a control.
2. Plan to add two drops of a drug-containing substance directly to the slide.
3. When you are finished testing one drug, you will need to flush the used *Daphnia* with the solution into a beaker of aged tap water provided by your teacher. Plan to use a new *Daphnia* for each substance tested.
4. ***Make sure your teacher has approved your experimental plan before you proceed further.***
5. Carry out your experiment. **CAUTION:** ***Wash your hands with soap and water immediately after making observations.***

DESIGN YOUR OWN BioLab

What drugs affect the heart rate of *Daphnia*?, *continued*

Chapter 36

Analyze and Conclude

1. **Making Inferences** Which drugs are stimulants? Which are depressants?

2. **Checking Your Hypotheses** Compare your predicted results with the experimental data. Explain whether or not your data support your hypotheses regarding the drugs' effects.

3. **Drawing Conclusions** How do the drugs affect the heart rate of this animal?

4. **Analyzing the Procedure** How would you alter your experiment if you did it again?

MiniLab 37-1

Experimenting

Checking Your Pulse

The heart speeds up when the blood volume reaching your right atrium increases. It also speeds up when the level of carbon dioxide in the blood rises. The number of heartbeats per minute is your heart rate, which can be measured by taking your pulse.

Procedure

1. Use the data table below.
2. Have a classmate take your resting pulse for 60 seconds while you are sitting at your lab table or desk. Use the photo on page 1013 of your text as a guide to finding your radial pulse.
3. Record your pulse in the table.
4. Repeat steps 2 and 3 four more times, then calculate your average resting pulse rate. Switch roles and take your classmate's resting pulse.
5. Exercise by doing "jumping jacks" for one minute.
6. Have your classmate take your pulse for 60 seconds immediately after exercising and record the value in the data table.
7. Repeat steps 5 and 6 four more times. Switch roles again with your classmate.

Data Table

Heart rate (beats per minute)		
Trial	**Resting**	**After exercise**
1		
2		
3		
4		
5		
Total		
Average		

MiniLab 37-1

Checking Your Pulse, *continued*

Analysis

1. Explain why your pulse is a means of indirectly measuring heart rate.

2. Use actual values from your data table to describe the changes that occur to your heart rate when exercising.

3. Suppose the amount of blood pumped by your left ventricle each time it contracts is 70 mL. Calculate your cardiac output (70 mL × heart rate per minute) while at rest and just after exercise.

Name Date Class

MiniLab 37-2

Experimenting

Testing Urine for Glucose

Glucose is a sugar that is needed by the body and is normally not present in the urine. When the concentration of glucose becomes too high in the blood, as happens with diabetes, glucose is filtered out by the kidneys.

Procedure

1. Use the data table.
2. Using a grease pencil, draw two circles on a glass slide. Mark one circle N, the other A.
3. Use a clean dropper to add two drops of "normal urine" to the circle marked N.
4. Use a clean dropper to add two drops of "abnormal urine" to the circle marked A.
5. Hold a small strip of glucose test paper in a forceps and touch it to the liquid in the drop labeled N. Remove it, wait 30 seconds, and record the color. A green color means glucose is present.
6. Use a new strip of glucose test paper to test drop A and record the color.
7. Test several unknown "urine" samples for the presence of glucose. Use a clean slide for each test.

Data Table

Urine sample	Color of test paper	Glucose present?
Normal (N)		
Abnormal (A)		
Unknown X		
Unknown Y		
Unknown Z		

Analysis

1. Which of the "unknown" samples could be from a person who has diabetes?

2. Which part of the test procedure could be considered your control? Explain your answer.

3. How could you explain your results if a test of normal urine indicated the presence of glucose?

Chapter 37

Measuring Respiration

Preparation

Problem

How can you measure respiratory rate and estimate tidal volume?

Objectives

In this BioLab, you will:

- **Measure** resting breathing rate.
- **Estimate** tidal volume by exhaling into a balloon.
- **Calculate** the amount of air inhaled per minute.

Materials

round balloon
string (1 m)
metric ruler
clock or watch with second hand

Skill Handbook

Use the **Skill Handbook** if you need additional help with this lab.

Procedure

Part A: Breathing Rate at Rest

1. Use Data Table 1.
2. Have your partner count the number of times you inhale in 30s.
3. Repeat step 2 two more times.
4. Calculate the average number of breaths.
5. Multiply the average number of breaths by two to get the average resting breathing rate in breaths per minute.

Data Table 1

Resting breathing rate	
Trial	**Inhalations in 30 sec**
1	
2	
3	
Average number breaths	
Breaths/min	

Part B: Estimating Tidal Volume

1. Use Data Table 2.
2. Take a regular breath and exhale normally into the balloon. Pinch the balloon closed.
3. Have a partner fit the string around the balloon at the widest part.
4. Measure the length of the string, in centimeters, around the circumference of the balloon. Record this measurement.
5. Repeat steps 2–4 four more times.
6. Calculate the average circumference of the five measurements.
7. Calculate the average radius of the balloon by dividing the average circumference by 6.28 (which is approximately equal to 2π).
8. Tidal volume is the amount of air expelled during a normal breath. Tidal volume can be determined using the balloon radius and the formula for determining the volume of a sphere.

$$\text{Volume} = \frac{4\pi r^3}{3}$$

where r = radius and $\pi = 3.14$. Calculate the average tidal volume using the average balloon radius.

Chapter 37

Measuring Respiration, *continued*

Data Table 2

Tidal volume	
Trial	**String measurement**
1	
2	
3	
4	
5	
Avg. circumference	
Avg. radius	
Avg. tidal volume	

9. Your calculated volume will be in cubic centimeters: 1 cm^3 = 1 mL.

Part C: Amount of Air Inhaled

1. Use Data Table 3.
2. Multiply the average tidal volume by the average number of breaths per minute to calculate the amount of air you inhale per minute.
3. Divide the number of milliliters of air by 1000 to get the number of liters of air you inhale per minute.

Data Table 3

Amount of air inhaled	
mL/min	
L/min	

Analyze and Conclude

1. **Making Comparisons** Compare your average number of breaths per minute and tidal volume per minute with those of other students.

2. **Thinking Critically** An average adult inhales 6000 mL of air per minute. Compare your estimated average volume of air with this figure. What factors could account for any differences?

3. **Making Predictions** Predict what would happen to your resting breathing rate after exercise.

MiniLab 38-1

Observing and Inferring

Examining Sperm and Egg Attraction

Most animals that reproduce by external fertilization live in water and release their sperm and eggs into the water. Somehow, a sperm and egg must meet. One adaptation of these animals that helps ensure fertilization is the release of thousands of sex cells at one time. The large number of sex cells increases the odds that at least some eggs will be fertilized. Chemical attraction could be another adaptation that encourages fertilization. If eggs give off a chemical that attracts sperm, each sperm would have help "finding" an egg.

Procedure

1 Place a dropperful of sea urchin eggs on a microscope slide. **CAUTION:** ***Use care when working with a microscope and microscope slides.***

2 While observing eggs under the microscope, add a drop of sea urchin sperm to the eggs.

Analysis

1. Describe the motion of a single sperm.

____

____

2. What cell structures are involved in providing energy for the sperm motion?

____

____

3. Are the sperm attracted to the eggs? How do you know?

____

____

Making and Using Graphs

MiniLab 38-2 Making a Graph of Fetal Size

You started out as a single cell. That cell divided by the process of mitosis to produce organ systems capable of maintaining an independent existence outside your mother's uterus. During the time you were in your mother's uterus, major changes took place. One of these changes involved your growth in length.

Table 38.1: Growth of a Fetus

Source of sample	Time after fertilization	Size
First trimester	3 weeks	3 mm
	4 weeks	6 mm
	6 weeks	12 mm
	7 weeks	2 cm
	8 weeks	4 cm
	9 weeks	5 cm
	3 months	7.5 cm
Second trimester	4 months	15 cm
	5 months	25 cm
	6 months	30 cm
Third trimester	7 months	35 cm
	8 months	40 cm
	9 months	51 cm

Procedure

1. Prepare a graph that plots time on the horizontal axis and length in centimeters on the vertical axis. Equally divide the horizontal axis into nine months. Then equally divide each of the first three months into four weeks.
2. Plot the data in *Table 38.1* on your graph.

Analysis

1. When is the fastest period of growth?

2. What structures are developing during this period of growth?

3. At what point does growth begin to slow down?

Name Date Class

Chapter 38

What hormone is produced by an embryo?

PREPARATION

Problem

How can you test for the presence of hCG?

Objectives

In this BioLab, you will:

- **Model** the chemicals used to test for the presence of hCG.
- **Interpret** the results of chemical reactions involving hCG in a pregnant and nonpregnant female.

Materials

scissors
heavy paper
tracing paper

Safety Precautions

Handle scissors with caution.

Skill Handbook

Use the **Skill Handbook** if you need additional help with this lab.

PROCEDURE

1. Use the data table below.
2. Copy models **A, B,** and **C** on page 1048 of your text onto tracing paper.
3. Copy the tracings onto heavy paper and cut them out. You will need 4 models of **A,** 4 models of **B,** and 1 model of **C.**
4. Model **A** represents a molecule of the hCG hormone. Model **B** represents a chemical called anti-hCG hormone. Model **C** represents a chemical that has four hCG molecules attached to it.
5. Note that the shapes of hCG and anti-hCG join together like puzzle pieces. These two chemicals react, or join together, when both are present in a solution. The shapes of anti-hCG and Chemical C also join, indicating that they chemically react when both are present. The combination of Chemical C and anti-hCG is green. Chemical C without anti-hCG attached is colorless.
6. Model the following events for the "Not pregnant" condition. Record them in the data table using drawings of the models.
   a. The hormone hCG is not present in the urine.
   b. Anti-hCG is added to a urine sample, then chemical C is added.
   c. Draw the resulting chemical in the data table and indicate the color that appears.

Data Table

Condition	hCG in urine?	+ Anti-hCG	= Joined hCG and anti-hCG?	+ Chemical C with anti-hCG?	Color
Not pregnant					
Pregnant					

What hormone is produced by an embryo?, *continued*

7. Model the following events for the "Pregnant" condition. Record them in the data table using drawings of the models.
   a. The hormone hCG is present.
   b. Anti-hCG is added to urine, then Chemical C is added.
   c. Draw the resulting chemical in the data table and indicate the color that appears.

Analyze and Conclude

1. **Analyzing** Explain the origin of hCG in a pregnant female.

2. **Analyzing** Explain why hCG is absent in a nonpregnant female.

3. **Concluding** Describe the roles of anti-hCG and Chemical C in both tests.

4. **Observing and Inferring** Explain why anti-hCG is added to the sample before Chemical C is added.

Name Date Class

MiniLab 39-1

Experimenting

Testing How Diseases are Spread

Microorganisms cannot travel over long distances by themselves. Unless they are somehow transferred from one animal or plant to another, infections will not spread. One method of transmission is by direct contact with an infected animal or plant.

Procedure

1. Label four plastic bags 1 to 4.
2. Put a fresh apple in bag 1 and seal the bag.
3. Rub a rotting apple over the entire surface of the remaining three apples. The rotting apple is your source of pathogens. **CAUTION:** ***Make sure to wash your hands after handling the rotting apple.***
4. Put one of the apples in bag 2.
5. Drop one apple to the floor from a height of about 2 m. Put this apple in bag 3.
6. Use a cotton ball to spread alcohol over the last apple. Let the apple air-dry and then place it in bag 4.
7. Store all of the bags in a dark place for one week.
8. Compare the apples and record your observations. **CAUTION:** ***Give all apples to your teacher for proper disposal.***

Analysis

1. What was the purpose of the fresh apple in bag 1?

____

____

2. Explain what happened to the rest of the apples.

____

____

3. Why is it important to clean a wound with disinfectant?

____

____

MiniLab 39-2

Observing and Inferring

Distinguishing Types of White Blood Cells

The human immune system includes five types of white blood cells found in the bloodstream: basophils, neutrophils, monocytes, eosinophils, and lymphocytes.

Procedure

1. Use the data table below.
2. Mount a prepared slide of blood cells on the microscope and focus on low power. Turn to high power and look for white blood cells. **CAUTION:** ***Use caution when working with microscope slides.***
3. Find a neutrophil, monocyte, eosinophil, and lymphocyte. You may see a basophil, although they are rare. Refer to the *Inside Story* for photos of these cells.
4. Count a total of 50 white blood cells, and record how many of each type you see.
5. Calculate the percentage by multiplying the number of each cell type by two. Record the percentages. Diagram each cell type.

Data Table

Type of white blood cell	Number counted	Percent	Diagram
Neutrophil			
Monocyte			
Basophil			
Lymphocyte			
Eosinophil			

1. Which type of white blood cell was most common? Second most common?

2. How do red and white blood cells differ?

Getting On-line for Information on Diseases

Chapter 39

PREPARATION

Problem

How can you use the Internet to obtain current research information on different diseases?

Objectives

In this BioLab, you will

- **Choose** five communicable and five noncommunicable diseases for study.
- **Use the Internet** to gather information.
- **Collect data** on the ten diseases and record it in a table.

Materials

access to the Internet

Skill Handbook

Use the **Skill Handbook** if you need additional help with this lab.

PROCEDURE

1. Use the two data tables.
2. Choose five communicable and five noncommunicable diseases you wish to investigate.
3. List the diseases in your data tables. Try to make your disease choices as specific as possible. For example, cancer as a topic is too broad. Instead, limit your choice to a specific type of cancer, such as breast cancer, prostate cancer, or Hodgkin's disease.
4. Go to the Glencoe Science Web Site at **www.glencoe.com/sec/science** to find helpful links to research information for this BioLab.
5. Be sure to complete the last two rows asking for current research findings and your sources of information.

Data Table 1

Communicable diseases					
	1	2	3	4	5
Disease name					
Organism responsible					
Classification of organism					
Mode of transmission					
Symptoms					
Treatment					
Current research					
Source of information					

INTERNET BioLab

Getting On-line for Information on Diseases, *continued*

Data Table 2

Noncommunicable diseases					
	1	**2**	**3**	**4**	**5**
Disease name					
Symptoms					
Organs affected					
Age group affected					
Treatment					
Current research					
Source of information					

Analyze and Conclude

1. **Defining** What is a pathogen? Provide several examples.

2. **Comparing** Describe the difference between a communicable and a noncommunicable disease. Provide several examples of each.

3. **Thinking Critically** What are vectors? Are they associated with communicable or noncommunicable diseases? Explain your answer.

4. **Applying Concepts** Explain why the table for noncommunicable diseases does not have a column for organism responsible or method of transmission.

5. **Using the Internet** What is one advantage of getting information on disease research by way of the Internet rather than from textbooks or an encyclopedia?

BioLab and MiniLab Worksheets

Answer Pages

Name Date Class

MiniLab 1-1 Predicting Whether Mildew Is Alive

Observing

What is mildew? Is it alive? We see it "growing" on plastic shower curtains or on bathroom grout. Does it show the characteristics associated with living things?

Procedure

1 Use the data table below.

Data Table

Prediction	Life characteristics
First	none
Second	
Third	

Expected Results: Students may initially predict that mildew is not alive. During microscopic examination, students will see long filaments (hyphae) and tiny circular objects (spores). Some students may see spores enclosed in a spore sac.

2 Predict whether or not mildew is alive. Record your prediction in the data table under "First Prediction."

3 Obtain a sample of mildew from your teacher. Examine it for life characteristics. Make a second prediction and record it in the data table along with any observed life characteristics. **CAUTION:** ***Wash hands thoroughly after handling the mildew sample.***

4 Following your teacher's directions, prepare a wet mount of mildew for viewing under the microscope. **CAUTION:** ***Use caution when working with a microscope, microscope slides, and coverslips.***

5 Are there any life characteristics visible through the microscope that you could not see before? Make a third prediction and include any observed life characteristics.

Analysis

1. Describe any life characteristics you observed.
   organization, reproduction, growth, adjusting to environment
2. Compare your three predictions and explain how your observations may have changed them.
   Answers will vary, depending on the observations made by individual students.
3. Explain the value of using scientific tools to extend your powers of observation.
   Certain life characteristics could not have been seen without the use of a microscope.

Name Date Class

MiniLab 1-2 Testing for Alcohol

Experimenting

Commercials for certain over-the-counter products may not tell you that one of the ingredients is alcohol. How can you verify whether or not a certain product contains alcohol? One way is to simply rely on the information provided during a commercial. Another way is to experiment and find out for yourself.

Procedure

1 Use the data table below.

Data Table

	Color of Liquid	Alcohol present
Circle A	**yellow-orange**	**No**
Circle B	**green to blue**	**Yes**
Circle C	**yellow**	**No**
Product name	**Data will vary.**	**Data will vary.**
Product name	**Data will vary.**	**Data will vary.**

2 Draw three circles on a glass slide. Label them A, B, and C. **CAUTION:** ***Put on safety goggles.***

3 Add one drop of water to circle A, one drop of alcohol to circle B, and one drop of alcohol-testing chemical to circles A, B, and C. **CAUTION:** ***Rinse immediately with water if testing chemical gets on skin or clothing.***

4 Wait 2–3 minutes. Note in the data table the color of each liquid and the presence or absence of alcohol.

5 Record the name of the first product to be tested.

6 Draw a circle on a clean glass slide. Add one drop of the product to the circle.

7 Add a drop of the alcohol-testing chemical to the circle. Wait 2–3 minutes. Record the color of the liquid.

8 Repeat steps 5–7 for each product to be tested. **CAUTION:** ***Wash your hands with soap and water immediately after using the alcohol-testing chemical.***

9 Complete the last column of the data table. If alcohol is present, the liquid turns green, deep green, or blue. A yellow or orange color means no alcohol is present.

Name Date Class

MiniLab 1-2

Testing for Alcohol, *continued*

Analysis

1. Explain the purpose of using the alcohol-testing chemical with water, with a known alcohol, and by itself.
   to determine which colors indicate the presence or absence of alcohol

2. Which products did contain alcohol? No alcohol?
   Answers will depend on products tested.

Name Date Class

MiniLab 1-3

Observing and Inferring

Hatching Dinosaurs

"Dinosaur eggs" can be found in specially marked packages of oatmeal. You will conduct an investigation to determine what causes these pretend eggs to hatch.

Procedure

Expected Results: The outer candy layer of a "dinosaur egg" melts away in boiling water in less than a minute, leaving a tiny dinosaur-shaped candy.

1. Use the data table below.

Data Table

	Before treatment	Hot water treatment	Cold water treatment
Appearance after one minute			

2. Observe the dinosaur eggs provided and record their characteristics in your table.
3. Place an egg in each of two containers.
4. Make a hypothesis about the water temperature that will cause the eggs to hatch.
5. Pour boiling water into one container and cold water in the other. Stir for one minute. Record your observations.

Analysis

1. Infer whether heat or moisture was more important for hatching eggs.
   heat or moisture

2. Design an experiment that would test either heat or moisture as the variable. What kind of quantitative data will you gather?
   To test heat, students might put eggs under a heat lamp. To test moisture, students might place eggs in various temperatures of water.

3. What will be your control?
   Variable is heat or moisture. Control is the basis of comparison for the tested condition. Students could measure "hatching" time.

4. How many trials will you run and how many eggs will you test? If time permits, conduct your experiment.
   Use as many eggs and trials as time permits.

Name Date Class

INTERNET BioLab

Collecting Biological Data

Chapter 1

Preparation

Problem
What life characteristics can be observed in a pill bug?

Objectives
In this BioLab, you will:
- **Observe** whether life characteristics are present in a pill bug.
- **Measure** the length of a pill bug.
- **Experiment** to determine if a pill bug responds to changes in its environment.
- **Use the Internet** to collect and compare data from other students.

Materials
pill bugs, *Armadillidium*
watch or classroom clock
container, glass or plastic
pencil with dull point
ruler
computer with Internet connection

Safety Precautions
Always wear goggles in the lab.

Skill Handbook
Use the **Skill Handbook** if you need additional help with this lab.

Procedure

1. Use the data table and graph outlines shown here.
2. Obtain a pill but from your teacher and place it in a small container.
3. Observe your pill bug to determine whether or not it has an orderly structure. Record your answer in the data table.
4. Using millimeters, measure and record the length of your pill bug in the data table.
5. Using your data and data from your classmates, complete the graph "Pill Bug Length: Classroom Data."
6. Go to the Glencoe Science Web Site at **www.glencoe.com/sec/science** to **post your data**.
7. *Gently* touch the underside of the pill bug with a *dull* pencil point. It may be necessary to gently flip the pill bug over with the pencil to get at its underside. **CAUTION:** ***Use care to avoid injuring the pill bug.***
8. Note its response and time, in seconds, how long the animal remains curled up. Record the time in the data table as Trial 1.
9. Repeat steps 7–8 four more times, recording each trial in the data table.
10. Calculate the average length of time your pill bug remains curled up in a ball.

Data Table

Organization and growth and development	
Orderly structure?	
Pill bug length in mm	

Response to environment	
Trial	Time in seconds
1	
2	
3	
4	
5	
Total	
Average time	

See Data and Observations on TWE p. 27.

Name Date Class

INTERNET BioLab

Collecting Biological Data, *continued*

Chapter 1

11. Post your data on the Glencoe Science Web Site.
12. Return the pill bug to your teacher. **CAUTION:** ***Wash your hands with soap and water after working with pill bugs.***

Pill Bug Length: Classroom Data
Number of Pill Bugs: 10 20 30 40 50 60 70 80 90 100
Length in Millimeters: 1 2 3 4 5 6 7 8 9 10 11 12 13 14 15 16 17 18 19 20

Pill Bug Length: Internet Data
Number of Pill Bugs: 1 2 3 4 5 6 7 8 9 10
Length in Millimeters: 1 2 3 4 5 6 7 8 9 10 11 12 13 14 15 16 17 18 19 20

Average Pill Bug Response Time: Internet Data
Number of Pill Bugs: 10 20 30 40 50 60 70 80 90 100
Response Time in Minutes: 1 2 3 4 5 6 7 8 9 10 11 12 13 14 15 16 17 18 19 20

Analyze and Conclude

1. **Thinking Critically** Explain how you would define the term "orderly structure." Explain how this trait might also pertain to nonliving things.
Student answers will vary and may include the following: shows organization; has specific parts for certain jobs; all pill bugs look alike. Yes, orderly structure also applies to nonliving things. Many nonliving things are organized, including buildings, books, computers.
2. **Using the Internet** Explain how data from the classroom and Internet graphs support the idea that pill bugs grow and develop.
The graph shows that pill bugs vary in length, but most reach a maximum size of about 10 mm.
3. **Interpreting Data** What was the most common length of time pill bugs remained curled in response to being touched?
Pill bugs remain curled for an average of approximately 20 seconds.
4. **Drawing a Conclusion** Explain how the response to being touched is an adaptation.
The pill bug's outer shell is rather tough; rolling into a ball could prevent predators from attacking its soft underside and fragile appendages.
5. **Experimenting** How might you design an experiment to determine whether or not pill bugs reproduce?
Student answers will vary. Place several pill bugs in a sealed container and add food and moisture as needed. Observe pill bugs weekly and compare numbers present to those originally placed in container. Or, look for immature forms that may appear in the container.

Name Date Class

MiniLab 2-1 Salt Tolerance of Seeds

Experimenting

Salinity, or the amount of salt dissolved in water, is an abiotic factor. Might salt water affect how certain seeds sprout or germinate? Experiment to find out.

Procedure

1. Soak 20 seeds in freshwater and 20 seeds in salt water overnight.
2. The next day, wrap the seeds in two different moist paper towels. Slide the towels into separate self-sealing plastic bags.
3. Label the bags "fresh" and "salt."
4. Examine all seeds two days later. Count the number of seeds in each treatment that show signs of root growth or sprouting, which is called germination. Record your data. **CAUTION:** ***Be sure to wash your hands after handling seeds.***

Analysis

1. Did the germination rates differ between treatments? If yes, how?
   Yes. Seeds soaked in tap water germinate. Seeds soaked in salt water do not.
2. What abiotic factor was tested in this experiment? What biotic factor was influenced?
   salinity of water; germination
3. Might all seeds respond to salt in a similar manner? How could you find out?
   No, each seed type may respond differently to salinity. Experimentation is needed.

Name Date Class

MiniLab 2-2 Detecting Carbon Dioxide

Observing and Inferring

Carbon dioxide is given off during respiration. When carbon dioxide is dissolved in water, an acid is formed. Certain chemicals called indicators can be used to detect acids. One indicator, called bromothymol blue, will change from its normal blue color to green or yellow if an acid is present.

Procedure

1. Half fill a test tube with bromothymol blue solution.
2. Add a quarter of an effervescent antacid tablet to the tube and note any color change.
3. Half fill a test tube with bromothymol blue solution. Using a straw, exhale into the bromothymol blue at least 30 times. **CAUTION:** ***Do not inhale the bromothymol blue.*** Record any color change in the test tube.

Analysis

1. Describe the color change that occurs when carbon dioxide is added to bromothymol blue.
   Bromothymol blue changes to green or yellow when carbon dioxide gas is added.
2. What was the chemical composition of the bubbles seen in the tube with the antacid tablet?
   carbon dioxide
3. Does exhaled air contain carbon dioxide? Explain.
   Yes, the color change in the bromothymol blue indicated the presence of carbon dioxide.

Name Date Class

DESIGN YOUR OWN BioLab

How can one population affect another?

Chapter 2

PREPARATION

Problem
How does a population of *Paramecium* react to a population of *Didinium*?

Hypotheses
Have your group agree on a hypothesis to be tested. Record your hypothesis.

Objectives
In this BioLab, you will:
- **Design** an experiment to establish the relationships between *Paramecium* and *Didinium*.
- **Use** appropriate variables, constants, and controls in experimental design.

Possible Materials
microscope
microscope slides
coverslips
culture of *Didinium*
culture of *Paramecium*
beakers or jars
eyedroppers
sterile pond water

Safety Precautions
Take care when using electrical equipment. Always use goggles in the lab. Handle slides and coverslips carefully. Dispose of broken glass in a container provided by your teacher.

Skill Handbook
Use the **Skill Handbook** if you need additional help with this lab.

PLAN THE EXPERIMENT

1. Review the discussion of feeding relationships in this chapter.
2. Decide which materials you will use in your investigation. Record your list.
3. Be sure that your experimental plan contains a control, tests a single variable such as population size, and allows for the collection of quantitative data.
4. Prepare a list of numbered directions. Explain how you will use each of your materials.

Check the Plan
Discuss the following points with other group members to decide final procedures. Make any needed changes to your plan.

1. What will you measure to determine the effect of the *Didinium* on *Paramecium*? If you count *Paramecia*, will you count all you can see in the field of vision of the microscope at a certain power? Will you have multiple trials? If so, how many?
2. What single factor will you vary? For example, will you put no *Didinium* in one culture of *Paramecium* and 5 mL of *Didinium* culture in another culture of *Paramecium*?
3. How long will you observe the populations?
4. How will you estimate the changes in the populations of *Paramecium* and *Didinium* during the experiment?
5. ***Your teacher must approve your plan before you proceed.***
6. Carry out your experiment.
7. Make a data table that has Date, Number of *Paramecium*, and Number of *Didinium* across the top. Place the data obtained for each culture in rows. Design and complete a graph of your data.

Name Date Class

DESIGN YOUR OWN BioLab

How can one population affect another?, *continued*

Chapter 2

ANALYZE AND CONCLUDE

1. **Analyzing Data** What differences did you observe among the experimental groups? Were these differences due to the presence of *Didinium*? Explain.
   Only cultures containing both *Didinium* and *Paramecium* showed a decline in numbers after a period of time. This difference was due to the presence of *Didinium* because they preyed upon the *Paramecium*.
2. **Drawing Conclusions** Did the *Paramecium* die out in any culture? Why or why not?
   Paramecium died out in the mixed culture. They were preyed upon by _Didinium_.
3. **Checking Your Hypothesis** Was your hypothesis supported by your data? If not, suggest a new hypothesis.
   In most cases, hypotheses will be supported.
4. **Thinking Critically** List several ways that your methods may have affected the outcome of the experiment.
   The list may include: counting errors, too few samples, or cultures becoming contaminated or being affected by temperature.

Data and Observations: Depending on the experiment, data and observations will vary. Typically, when the populations are mixed together, both will initially increase in number. A decrease in prey will then be detected as the predator population feeds upon them, with a final drop in the population of predators as their food supply runs out.

Name Date Class

MiniLab 3-1 Looking at Lichens

Observing

Lichens have the reputation for being a pioneer species when it comes to succession. They often inhabit rocky areas and start the process of soil formation. How is it possible for lichens to grow on rock?

Procedure

1. Examine the lichen samples provided by your teacher. Note their color, shape, and texture.
2. Use a microscope to examine a prepared slide of a stained section of a lichen. Use low-power magnification and then change to high power as needed.
3. Observe the dark bodies that are cells containing chloroplasts. Notice that lichens are composed of an alga and a fungus. Diagram what you see.

Analysis

1. Describe the general appearance of a whole lichen and of the lichen under a microscope.
   Color may be dull green, red, orange, or yellow; shape may be crusty or flat. Fungus portion may be long, clear strands; alga portion, small green cells.
2. How does a lichen illustrate mutualism?
   The fungus receives food from the alga, and the alga receives moisture from the fungus.
3. Explain how mutualism explains why lichens are able to survive on rocks.
   Rocks offer harsh living conditions. The algae and fungi in lichens overcome this by making their own food and retaining moisture.

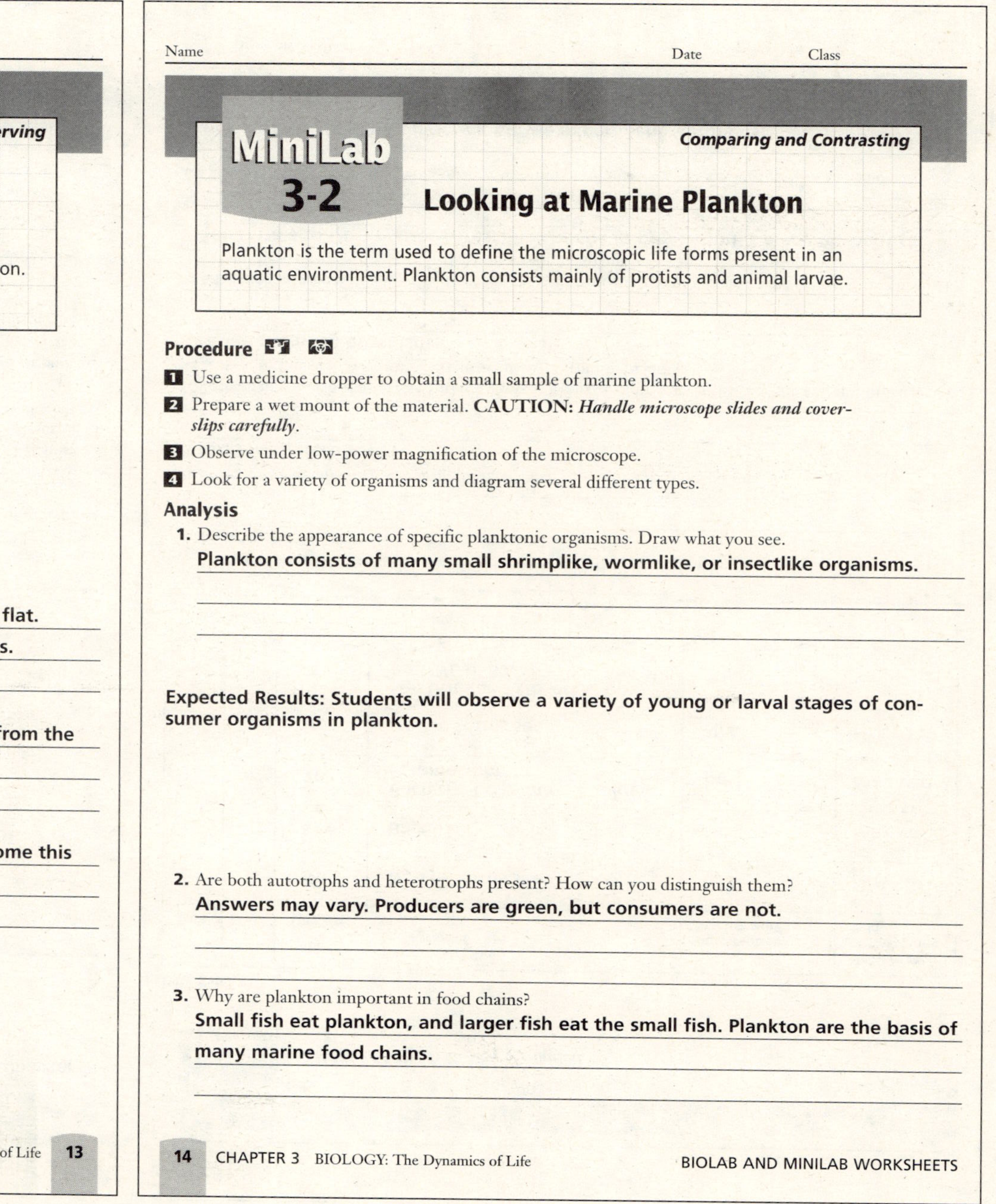

Name Date Class

MiniLab 3-2 Looking at Marine Plankton

Comparing and Contrasting

Plankton is the term used to define the microscopic life forms present in an aquatic environment. Plankton consists mainly of protists and animal larvae.

Procedure

1. Use a medicine dropper to obtain a small sample of marine plankton.
2. Prepare a wet mount of the material. **CAUTION:** ***Handle microscope slides and coverslips carefully.***
3. Observe under low-power magnification of the microscope.
4. Look for a variety of organisms and diagram several different types.

Analysis

1. Describe the appearance of specific planktonic organisms. Draw what you see.
   Plankton consists of many small shrimplike, wormlike, or insectlike organisms.

Expected Results: Students will observe a variety of young or larval stages of consumer organisms in plankton.

2. Are both autotrophs and heterotrophs present? How can you distinguish them?
   Answers may vary. Producers are green, but consumers are not.
3. Why are plankton important in food chains?
   Small fish eat plankton, and larger fish eat the small fish. Plankton are the basis of many marine food chains.

Name Date Class

INVESTIGATE BioLab

Succession in a Jar

Chapter 3

Preparation

Problem
Can you observe succession in a pond water ecosystem?

Objectives
In this BioLab, you will:
- **Observe** changes in three pond water environments.
- **Count** the number of each type of organism seen.
- **Determine** if the changes observed illustrate succession.

Materials
glass jars, 3
pasteurized spring water
pond water containing plant material
labels
glass slides and cover glasses
droppers
plastic wrap
cooked white rice
teaspoon, plastic
microscope

Safety Precautions
Always wear goggles in the lab.

Skill Handbook
Use the **Skill Handbook** if you need additional help with this lab.

Procedure

1. Examine the pond water sample provided by your teacher.
2. Fill three glass jars with equal amounts of pasteurized (sterilized) spring water.
3. Label them A, B, and C. Add your name and today's date to each label.
4. Add the following to each of your three jars:
   Jar A: Nothing else
   Jar B: 3 grains of cooked white rice
   Jar C: 3 grains of cooked white rice, one teaspoon of pond sediment, and a small amount of any plant material present in the pond water.
5. Record the turbidity of each jar in the data table below. Turbidity means cloudiness and can best be judged by comparing jar A to B to C. Score turbidity on a scale of 1 to 10, with 1 meaning very clear water and 10 meaning very cloudy water.
6. Gently swirl the contents of each jar.
7. Using a different clean dropper for each jar, remove a sample and prepare a wet mount of the liquid from each jar. Label each glass slide A, B, or C to avoid any mix-up. CAUTION: *Handle glass slides, coverslips, and glassware carefully.*
8. Observe each slide under low power. Look for autotrophic and heterotrophic organisms. Identify these organisms by name, describe their appearance, or make a sketch of what they look like.
9. Report the number of each type of organism as viewed in a low power field of view (the circle of light seen when looking through the microscope under low power is your field of view).
10. Complete the data table for your first observations.
11. Cover each jar with either a lid or plastic wrap and place them in a lighted area.
12. Observe the jars every three days for several weeks, repeating steps 5–11 each time an observation is made. Use your data table to record your observations.

Name Date Class

INVESTIGATE BioLab

Succession in a Jar, *continued*

Chapter 3

Data Table

Date	Jar	Turbidity	Name, description, or diagram of organism seen	Autotroph or heterotroph?	Number seen per low power field
	A	No	No life forms seen	N/A	N/A
	B	Yes	Bacteria	No	May vary
	C	Yes	Bacteria, *Colpidium, Euplotes, Paramecium*	Both	May vary
	A	No	No life forms seen	N/A	N/A
	B	—	See Data and Observations on TWE p. 89.	—	—

Analyze and Conclude

1. **Applying Concepts** Which of the three jars was a control? Explain why.
   A; only pasteurized water was added.
2. **Observing and Inferring** What might have been the role of the cooked rice?
   food for heterotrophs
3. **Recognizing Cause and Effect** Turbidity was a means of indirectly measuring the amount of bacterial growth in the jars. Why was there little, if any, turbidity in jar A?
   It had no living organisms.
4. **Analyzing Information** Describe the changes over time in the number and type of heterotrophs. Could these changes be described as succession? Explain.
   See Data and Observations. Yes; any change in population is a key element of succession.
5. **Observing and Inferring** Were you able to observe a climax ecosystem during this experiment? Explain your answer.
   Answers will depend on how long the experiment continues.

Name Date Class

MiniLab 4-1

Making and Using Tables

Fruit Fly Population Growth

Fruit Flies (*Drosophila melanogaster*) and similar insects have rapid rates of reproduction. Fruit flies are frequently used in biological research because they reproduce quickly and are easy to keep and count. In this activity, you will observe the growth of a fruit fly population as it exploits a food supply.

Procedure

1. Place half a banana in a jar and allow it to sit outside in a warm shaded area, or in a warm area in your classroom.
2. Leave the jar for one day or until you have at least three fruit flies in it. Put the mesh on top of the jar and fasten with the rubber band.
3. Each day record how many adult fruit flies are alive in the jar. Record for at least three weeks. Put your data into table form. **CAUTION:** ***Return the fruit flies to your teacher for proper disposal.***

Analysis

1. How many fruit flies did you start with? On what day were there the most fruit flies? How many were there?
   Answers will vary.
2. Why did the number of fruit flies decrease?
   lack of food due to overcrowding
3. Based on this investigation, why are insects considered to display a rapid reproduction pattern?
   The population quickly increased and then quickly decreased.

Expected Results: Fruit flies will arrive and quickly reproduce. Larvae will be evident on the jar and the banana. In a few weeks, the jar will be full of flies. Soon the population will decline because the food supply is limited.

Name Date Class

MiniLab 4-2

Using Numbers

Doubling Time

The time needed for any population to double its size is known as its "doubling time." For example, if a population grows slowly, its doubling time will be long. If it is growing rapidly, its doubling time will be short.

Procedure

1. The following formula is used to calculate a population's doubling time:

$$\text{Doubling time (in years)} = \frac{70}{\text{annual percent growth rate}}$$

2. Use the data table below.
3. Complete the table by calculating the doubling time of human populations for the listed geographic regions.

Data Table

Geographic region	Annual percent growth rate	Doubling time
Africa	2.8	**25 years**
Latin America	2.2	**31.8 years**
Asia	1.9	**36.8 years**
North America	0.7	**100 years**
Europe	0.3	**233 years**

Analysis

1. Which region has the fastest doubling time? Slowest doubling time?
   Africa; Europe
2. How might this type of information be useful to governments of these regions?
   Answers may vary, but students may mention the planning of roads, homes, and other needed structures.
3. What are some of the ecological implications for an area with a fast doubling time?
   Students may list the high potential for disease, problem with waste disposal, and lack of space.

Name Date Class

INVESTIGATE BioLab

How can you determine the size of an animal population?

Chapter 4

Preparation

Problem

How can you model a measuring technique to determine the size of an animal population?

Objectives

In this BioLab you will:

- **Model** the procedure used to measure an animal population.
- **Collect** data on a modeled animal population.
- **Calculate** the size of a modeled animal population.

Materials

paper bag containing beans
magic marker
calculator (optional)

Safety Precautions

Always wear goggles in the lab. Wash hands after working with plant material.

Skill Handbook

Use the **Skill Handbook** if you need additional help with this lab.

Procedure

1. Use the data table below.
2. Reach into your bag and remove 20 beans.
3. Use a dark magic marker to color these beans. These will represent your caught and marked animals.
4. When the ink has dried, return the beans to the bag.
5. Shake the bag. Without looking into the bag, reach in and remove 30 beans.
6. Record the number of marked beans (recaught and marked) and the number of unmarked beans (caught and unmarked) in your data table as trial 1.
7. Return all the beans to the bag.
8. Repeat steps 5 to 7 four more times for trials 2 to 5.
9. Calculate averages for each of the columns.
10. Using average values, calculate the original size of the bean population in the bag by using the following formula:
    M = number initially marked
    CwM = average number caught during the trials with marks
    Cw/oM = average number caught during the trials without marks

$$\text{Calculated Population Size} = \frac{M \times (CwM + Cw/oM)}{CwM}$$

11. Record the calculated population size in the data table.
12. To verify the actual population size, count the total number of beans in the bag and record this value in the data table.

Name Date Class

INTERNET BioLab

How can you determine the size of an animal population?, *continued*

Chapter 4

Data Table

Trial	Total caught	Number caught with marks	Number caught without marks
1	20	3	30
2	20	2	28
3	20	5	28
4	20	4	30
5	20	6	28
Totals	100	20	144
Averages	20	4	28.8

Calculated population size = **164**

Actual population size = **175**

Analyze and Conclude

1. **Thinking Critically** This experiment is a simulation. Explain why this type of activity is best done as a simulation.
   Students may note that it's difficult to obtain enough traps, find a suitable site for study, and work with trapped animals.
2. **Applying Concepts** Give an example of how this technique could actually be used by a scientist.
   Possible answers: to determine the size of a population in a specific area, such as a national park, or to measure a change in population size over time
3. **Analyzing Data** Compare the calculated to the actual population size. Explain why they may not agree exactly. What changes to the procedure would improve the accuracy of the activity?
   The calculated value is based on counting a representation of the population, not every individual in the population. To increase the accuracy of the activity, one could increase the number of trials, the number of the animals recaught, or the number of animals caught and marked.
4. **Making Inferences** Explain why this technique is used more often with animals than with plants when calculating population size.
   Plants do not move about and so are easier to count.
5. **Making Predictions** Assume you were doing this experiment with living animals. What would you be doing in step 2? Step 3? Step 5?
   Step 2, capture animals in cages; step 3, tag or mark the animals in some manner; step 5, reset traps and capture more animals

Name Date Class

MiniLab 5-1

Using Numbers

Measuring Species Diversity

Index of diversity (I.D.) is a mathematical way of expressing the amount of biodiversity and species distribution in a community. Communities with many different species (a high index of diversity) will be able to withstand environmental changes better than communities with only a few species (a low index of diversity.)

Procedure

1. Use the data table below.
2. Walk a city block or an area designated by your teacher and record the number of *different species* of trees present (you don't have to know their names, just that they differ by species). Record this number in your data table.
3. Walk the block or area again. This time, make a list of the trees by assigning each a number as you walk by it. Place an X under Tree 1 on your list. If Tree 2 is the same species as Tree 1, mark an X below it. Continue to mark an X under the trees as long as the species is the same as the previous one. When a different species is observed, mark an O under that tree on your list. Continue to mark an O if the next tree is the same species as the previous. If the next tree is different, mark an X.
4. Record in your data table:
   - **a.** the number of "runs." Runs are represented by a group of similar symbols in a row. Example—XXXXOOXO would be four runs (XXXX = 1 run, OO = 2 runs, X = 3, O = 4).
   - **b.** the total number of trees counted.
5. Calculate the Index of Diversity (I.D.) using the formula in the data table.

Data Table
Number of species =
Number of runs =
Number of trees =
Index of diversity = $\frac{\text{Number of species} \times \text{number of runs}}{\text{Number of trees}}$ =

Expected Results:

Numbers below one (decimals) indicate very low species diversity. Numbers over 1 show better diversity. A typical city street may yield an I.D. of slightly over 1.

Name Date Class

MiniLab 5-1

Measuring Species Diversity, *continued*

Analysis

1. Compare how your tree I.D. might compare with that of a vacant lot and with that of a grass lawn. Explain your answer.
   A vacant lot might have a higher I.D. because many different species might be present in a small area. A grass lawn would have a much lower I.D., perhaps with only one species present.
2. If humans were concerned about biological diversity, would it be best to have a low or high I.D. for a particular environment? Explain your answer.
   A higher I.D. would indicate greater species diversity. Communities with a low I.D. may be prone to species extinction if environmental conditions change.

Name Date Class

MiniLab 5-2

Experimenting

Conservation of Soil

Soil is an important natural resource and should be conserved. How does one conserve soil? As a start, we should be aware of factors that promote or speed up its unnecessary loss or erosion.

Procedure

1. Use the data table below.

Data Table

Source of sample	Volume of original water	Volume of collected water	Volume of eroded soil
Bare soil			
Soil with grass			

2. Measure 200 mL of water in a beaker.
3. Fill a plastic tray with soil as shown on page 126 of your text.
4. Pour the water onto the soil, tilting the tray and placing a dish underneath the end of it as indicated in the diagram on page 126.
5. Wait several minutes for all water and soil to drain into the dish.
6. Pour the soil and water from the dish into a graduated cylinder. Wait several minutes for the soil and water to settle. Measure the volume of soil and water that washed or eroded into the dish. Record these values in your data table.
7. Repeat steps 2–6, only this time place a section of soil in which grass is growing onto the tray. **CAUTION:** ***Always wash your hands with soap and water after working with soil.***

Analysis

1. What part of the experiment simulated soil erosion?

   water and soil draining into the dish

2. Based on this experiment, explain why farmers usually plant unused fields with some type of crop cover.

   to reduce soil erosion

Expected Results: There will be less measured soil erosion when vegetation is present than when soil is bare.

Name Date Class

INTERNET BioLab Researching Information on Exotic Pets

Chapter 5

Preparation

Problem

How can you use the Internet to gather information on keeping an exotic animal as a pet?

Objectives

In this BioLab, you will:

- **Select** on animal that is considered an exotic pet.
- **Use the Internet** to collect and compare information from other students.
- **Conclude** if the animal you have chosen would or would not make a good pet.

Materials

access to the Internet

Procedure

1. Use the data table as a guide for the information to be researched.
2. Pick an exotic pet from the following list of choices: hedgehog, snake, ferret, large cat such as a tiger or panther, monkey, ape, iguana, tropical bird.
3. Go to the Glencoe Science Web Site at **www.glencoe.com/sec/science to find links** that will provide you with information for this BioLab. Note: You are not limited to the pet suggestions provided. If a different organism appeals to you, research it instead.
4. Record your findings in the data table.

Data Table

Category	Response
Exotic pet choice	
Scientific name	
Natural habitat (where found in nature)	
Adult size	
Dietary needs	
Special health problems	
Source of medical care, if needed	
Safety issues for humans	
Size of cage area needed	
Special environmental needs	
Social needs	
Cost of purchase	
Cost of maintaining (monthly estimate)	
Care issues (high/low maintenance)	
Additional information	
Additional sources	

Name Date Class

INTERNET BioLab Researching Information on Exotic Pets, *continued*

Chapter 5

Analyze and Conclude

Analyze and Conclude

1. **Defining Operationally** What is meant by the term *domesticated*?
   Domesticated **means bred for an extended period of time in captivity.**
2. **Using the Internet** Look at the findings posted by other students. Which of the animals researched would make the best pet? Which would not be a wise choice? Explain.
   Answers will depend on the information students located.
3. **Interpreting Data** What do you consider the most important information gained from your research that:
   a. supports keeping your exotic pet choice?
      Answers may include that exotic pets are interesting, have a good market for resale, attract attention, and are challenging to raise.
   b. does not support keeping your exotic pet choice?
      Answers may include that these animals are expensive and difficult to feed and care for, including finding veterinarian services.
4. **Thinking Critically** What positive contribution might be made to the cause of conservation when keeping an exotic pet? Explain.
   Answers may include that keeping these pets allows for possible breeding, thus helping to save an endangered species.
5. **Thinking Critically** How can keeping exotic pets be a negative influence on conservation biology efforts?
   Students may note that buying these pets endangers species even more by encouraging black-market trade.
6. **Analyzing Information** What are some reasons why zoos rather than individuals are better able to handle exotic animals?
   Zoos can better meet the animals' needs for food, shelter, and medical care.

Data and Conclusions: Student data and conclusions will differ, depending on the animals they researched, the web sites they visited, and other references they used.

Name Date Class

Experimenting

MiniLab 6-1 Determine pH

The pH of a solution is a measurement of how acidic or basic that solution is. An easy way to measure the pH of a solution is to use pH paper.

Procedure

1. Pour a small amount (about 5 mL) of each of the following into separate clean, small beakers or other small glass containers: lemon juice, household ammonia solution, liquid detergent, shampoo, and vinegar.
2. Dip a fresh strip of pH paper briefly into each solution and remove.
3. Compare the color of the wet paper with the pH color chart; record the pH of each material. **CAUTION:** ***Wash your hands after handling lab materials.***

Analysis

1. Which solutions are acids?
lemon juice and vinegar

2. Which solutions are bases?
household ammonia and liquid detergent

3. What ions in the solution caused the pH paper to change? Which solution contained the highest concentration of hydroxide ions? How do you know?
H^+ ions and OH^- ions. Household ammonia contains the most OH^- ions; it had the highest pH.

Expected Results: The approximate pH of the solutions are: lemon juice, pH 3; household ammonia, pH 11; liquid detergent, pH 10; shampoo, pH 7; and vinegar, pH 3.

Name Date Class

Applying Concepts

MiniLab 6-2 Examine the Rate of Diffusion

In this lab, you will place a small potato cube in a solution of potassium permanganate and observe how far the dark purple color diffuses into the potato after a given length of time.

Procedure

1. Using a single-edge razor blade, cut a cube 1 cm on each side from a raw, peeled potato. **CAUTION:** ***Be careful with sharp objects.*** **Do not cut objects while holding them in your hand.**
2. Place the cube in a cup or beaker containing the purple solution. The solution should cover the cube. Note and record the time. Let the cube stand in the solution for between 10 and 30 minutes.
3. Using forceps, remove the cube from the solution and note the time. Cut the cube in half.
4. Measure, in millimeters, how far the purple solution has diffused, and divide this number by the time you allowed your potato to remain in the solution. This is the diffusion rate.

Analysis

1. How far did the purple solution diffuse?
Answers will depend on the amount of time the cube is in the solution.

2. What was the rate of diffusion per minute?
The rate will be in the tenths to hundredths of millimeters per minute.

Expected Results: The color will diffuse only a few millimeters into the cube, the exact distance depending upon the amount of time it is in the solution.

Name Date Class

DESIGN YOUR OWN BioLab

Does temperature affect an enzyme reaction?

Chapter 6

Preparation

Problem

Does the enzyme peroxidase work in cold temperatures? Does peroxidase work better at higher temperatures? Does peroxidase work after being frozen or boiled?

Hypotheses

Make a hypothesis regarding how you think temperature will affect the rate at which the enzyme peroxidase breaks down hydrogen peroxide. Consider both low and high temperatures.

Objectives

In this BioLab, you will:

- **Observe** the activity of an enzyme.
- **Compare** the activity of the enzyme at various temperatures.

Possible Materials

clock or timer
400-mL beaker
kitchen knife
tongs or large forceps
5-mm thick potato slices
3% hydrogen peroxide
ice
hot plate
waxed paper
thermometer

Safety Precautions

Be sure to wash your hands before and after handling the lab materials. Always wear goggles in the lab.

Skill Handbook

Use the **Skill Handbook** if you need additional help with this lab.

Plan the Experiment

1. Decide on a way to test your group's hypothesis. Keep the available materials in mind.
2. When testing the activity of the enzyme at a certain temperature, consider the length of time it will take for the potato to reach that temperature, and how the temperature will be measured.
3. To test for peroxidase activity, add 1 drop of hydrogen peroxide to the potato slice and observe what happens.
4. When heating a thin potato slice, first place it in a small amount of water in a beaker. Then heat the beaker slowly so that the temperature of the water and the temperature of the slice are always the same. Try to make observations at several temperatures between 10°C and 100°C.

Check the Plan

Discuss the following points with other groups to decide on the final procedure for your experiment.

1. What data will you collect? How will you record them?
2. What factors should be controlled?
3. What temperatures will you test?
4. How will you achieve those temperatures?
5. *Make sure your teacher has approved your experimental plan before you proceed further.*
6. Carry out your experiment. **CAUTION:** ***Be careful with chemicals and heat. Wash your hands after the lab.***

Name Date Class

DESIGN YOUR OWN BioLab

Does temperature affect an enzyme reaction?, *continued*

Chapter 6

Analyze and Conclude

Analyze and Conclude

1. **Checking Your Hypothesis** Do your data support or reject your hypothesis? Explain your results.
   Students should explain whether their data support or reject their hypotheses.
2. **Analyzing Data** At what temperature did peroxidase work best?
   Between 20°C–50°C
3. **Identifying Variables** What factors did you need to control in your tests?
   Answers may include the amount of time each potato was exposed to the temperature, the sizes of the potato slices, and the amount of peroxide added.
4. **Recognizing Cause and Effect** If you've ever used hydrogen peroxide as an antiseptic to treat a cut or scrape, you know that it foams as soon as it touches an open wound. How can you account for this observation?
   Human tissue contains peroxidase, so the hydrogen peroxide is broken down and releases oxygen.

Data and Observations: Cooling will not deactivate the enzymes but can slow the overall reaction. Potato slices heated over 70° will not generate oxygen bubbles.

Name Date Class

MiniLab 7-1 Measuring Objects Under a Microscope

Measuring in SI

Knowing the diameter of the circle of light you see when looking through a microscope allows you to measure the size of objects that are being viewed. For most microscopes, the diameter of the circle of light is 1.5 mm, or 1500 µm (micrometers), under low power and 0.375 mm, or 375 µm, under high power.

Refer to *Practicing Scientific Methods* in the **Skill Handbook** if you need help with SI units.

Procedure

1. Look at diagram A on page 177 of your text that shows an object viewed under low power. Knowing the circle diameter to be 1500 µm, the estimated length of object (a) is 400 µm. What is the estimated length of object (b)?
   700 µm
2. Look at diagram B on page 177 of your text that shows an object viewed under high power. Knowing the circle diameter to be 375 µm, the estimated length of object (c) is 100 µm. What is the estimated length of object (d)?
   25 µm
3. Prepare a wet mount of a strand of your hair. Your teacher can help with this procedure. CAUTION: ***Use caution when handling microscopes and glass slides.*** Measure the width of your hair strand while viewing it under low and then high power.
   60–100 µm (depending on race and/or hair color)

Analysis

1. An object can be magnified 100, 200, or 1000 times when viewed under a microscope. Does the object's actual size change with each magnification? Explain.
   No. An object's size does not change, only its magnification changes.
2. Do your observations of the size of your hair strand under low and high power support the answer to question 1? If not, offer a possible explanation why.
   Answers will vary, but observations under low and high power should be close. Error in estimating may be the cause of differing measurements.

Name Date Class

MiniLab 7-2 Cell Organelles

Experimenting

Adding stains to cellular material helps you distinguish cell organelles.

Procedure

CAUTION: ***Be sure to wash hands before and after this experiment.***

1. Prepare a water wet mount of onion skin. Do this by using your fingernail to peel off the inside of a layer of onion bulb. The layer must be almost transparent. Use the diagram on page 187 of your text as a guide.
2. Make sure that the onion layer is lying flat on the glass slide and not folded.
3. Observe the onion cells under low- and high-power magnification. Identify as many organelles as possible.
4. Repeat steps 1 through 3, only this time use an iodine stain instead of water.

Analysis

1. What organelles were easily seen in the unstained onion cells? Cells stained with iodine?
   cell wall, cytoplasm; cell wall, cytoplasm, nucleus
2. How are stains useful for viewing cells?
   Staining allows certain organelles to be more easily observed.

Name Date Class

INVESTIGATE BioLab

Observing and Comparing Different Cell Types

Chapter 7

Preparation

Problem
Are all cells alike in appearance and size?

Objectives
In this BioLab, you will:
- **Observe**, **diagram**, and **measure** cells and their organelles.
- **Hypothesize** which cells are from prokaryotes, eukaryotes, unicellular organisms, and multicellular organisms.
- **List** the traits of plant and animal cells.

Materials
microscope
glass slide
water
dropper
coverslip
forceps
prepared slides of *Bacillus subtilis*, frog blood, and *Elodea*

Safety Precautions
Always wear goggles in the lab.

Skill Handbook
Use the **Skill Handbook** if you need additional help with this lab.

Procedure

1. Use the data table below.
2. Examine a prepared slide of *Bacillus subtilis* using both low- and high-power magnification. (NOTE: This slide has been stained. Its natural color is clear.)
CAUTION: ***Use care when handling slides. Dispose of any broken glass in a container provided by your teacher.***

Data Table

	Bacillus subtilis	*Elodea*	Frog Blood
Organelles observed	No visible internal organelles	Chloroplasts, nucleus, cell wall	Nucleus, plasma membrane
Prokaryote or eukaryote	Prokaryote	Eukaryote	Eukaryote
From a multicellular or unicellular organism	Unicellular	Multicellular	Multicellular
Diagram (with size in micrometers, μm)	10–20 μm	40–80 μm	40–50 μm

3. Look for and record the names of any observed organelles. Hypothesize if these cells are prokaryotes or eukaryotes. Hypothesize if these cells are from a unicellular or multicellular organism. Record your findings in the table.
4. Diagram one cell as seen under high-power magnification.
5. While using high power, determine the length and width in micrometers of this cell. Refer to *Practicing Scientific Methods* in the **Skill Handbook** for help with determining magnification. Record your measurements on the diagram.
6. Prepare a wet mount of a single leaf from *Elodea* using the diagram as a guide.
7. Observe cells under low- and high-power magnification.
8. Repeat steps 3 through 5 for *Elodea*.
9. Examine a prepared slide of frog blood. (NOTE: This slide has been stained. Its natural color is pink.)
10. Observe cells under low- and high-power magnification.
11. Repeat steps 3 through 5 for frog blood cells.

Analyze and Conclude

1. **Observing and Inferring** Which cells were prokaryotes? How were you able to tell?
***Bacillus subtilis* was a prokaryote—no internal organelles were visible.**
2. **Observing and Inferring** Which cells were eukaryotes? How were you able to tell?
***Elodea* and epithelium cells were eukaryotic—internal organelles were observed.**
3. **Predicting** Which cell was from a plant, from an animal? Explain your answer.
***Elodea* was from a plant—cell wall and chloroplasts were visible. Blood cells were from an animal—no cell wall or chloroplasts were visible.**
4. **Measuring** Are prokaryote or eukaryote cells larger? Give specific measurements to support your answer.
Eukaryote cells are larger—the prokaryotic cell was only about 10 μm while both eukaryotic cells were in the range of 50–80 μm.
5. **Defining Operationally** Describe how plant and animal cells are alike and how they differ.
Both plant and animal cells are eukaryotic, both have organelles such as plasma membranes, nucleus, and cytoplasm. Plant cells have cell walls and chloroplasts.

Name Date Class

MiniLab 8-1 Cell Membrane Simulation

Formulating Models

If membranes show selective permeability, what might happen if a plastic bag (representing a cell's membrane) were filled with starch molecules on the inside and surrounded by iodine molecules on the outside?

Procedure

1. Fill a plastic bag with 50 mL of starch. Seal the bag with a twist tie.
2. Fill a beaker with 50 mL of of iodine solution. **CAUTION:** ***Rinse with water if iodine gets on skin. Iodine is toxic.***
3. Note and record the color of the starch and iodine.
4. Place the bag into the beaker. **CAUTION:** ***Wash your hands after handling lab materials.***
5. Note and record the color of the starch and iodine 24 hours later.

Analysis

1. Describe and compare the color of the iodine and starch at the start and at the conclusion of the experiment.

Start—starch was clear, iodine was rust; end—starch was purple, iodine was rust.

2. Fact: Starch mixed with iodine forms a purple color.

a. In which direction did the iodine move? What is your evidence?

Iodine moved into the bag as shown by the purple color.

b. In which direction did the starch move? What is your evidence?

Starch did not move out of the bag as shown by no color change outside of the bag.

3. Explain how this experiment illustrates selective permeability.

The plastic membrane let iodine pass into the bag but blocked starch from passing out.

Expected Results: The inside of the bag will be purple indicating passage of iodine into the bag. The outside of the bag will be rust color, indicating starch did not pass out of the bag.

Name Date Class

MiniLab 8-2 Seeing Asters

Comparing and Contrasting

The result of the process of mitosis is similar in plant and animal cells. However, animal cells have asters whereas plant cells do not. Animal cells undergoing mitosis clearly show these structures.

Procedure

1. Examine a slide marked "fish mitosis" under low- and high-power magnification. **CAUTION:** ***Use care when handling prepared slides.***
2. Find cells that are undergoing mitosis. You will be able to see dark-stained rodlike structures within certain cells. These structures are chromosomes.
3. Note the appearance and location of asters. They will appear as ray or starlike structures at opposite ends of cells that are in metaphase.
4. Asters may also be observed in cells that are in other phases of mitosis.

Analysis

1. Describe the appearance and location of asters in cells that are in prophase.

Asters are starlike projections of microtubules associated with centrioles. Asters are found at the cell poles in prophase.

2. Explain how you know that asters are not critical to mitosis.

Asters are not critical because plant cells undergo mitosis without the structures.

3. Design an experiment that tests the hypothesis that asters are not essential for mitosis in animal cells.

Laser microbeams can be aimed to destroy specific organelles. Destroy the asters in a dividing cell and note if the cell completes mitosis.

INVESTIGATE BioLab

Where is mitosis most common?

Chapter 8

Preparation

Problem
Does mitosis occur at the same rate in all parts of an onion root?

Objectives
In this BioLab, you will:
- **Observe** cells in two different root areas.
- **Identify** the stages of mitosis in each area.

Materials
prepared slide of onion root tip
microscope

Skill Handbook
Use the **Skill Handbook** if you need additional help with this lab.

Procedure

1. Use the data table below.
2. Using Diagram **A** on page 221 of your text as a guide, locate area X on a prepared slide of onion root tip.
3. Place the prepared slide under your microscope and use low power to locate area X. **CAUTION:** ***Use care when handling prepared slides.***
4. Switch to high power.
5. Using Diagram **B** on page 221 of your text as a guide:
   a. Identify those cells that are in mitosis and in interphase.
   b. Record in the data table the number of cells observed in each phase of mitosis and interphase for area X.

   (NOTE: It will be easier to count and keep track of cells by following rows. See Diagram C on page 221 of your text as a guide to counting.)
6. Using Diagram **A** again, locate area Y on the same prepared slide.
7. Place the prepared slide under your microscope and use low power to locate area Y.
8. Switch to high power.
9. Using Diagram **B** as a guide:
   a. Identify those cells that are in mitosis and in interphase.
   b. Record in the data table the number of cells observed in each phase of mitosis and interphase for area Y.

Data Table Sample data

Phase	Area X	Area Y
Interphase	95	36
Prophase	15	0
Metaphase	5	0
Anaphase	2	0
Telophase	4	0

INVESTIGATE BioLab

Where is mitosis most common?, *continued*

Chapter 8

Analyze and Conclude

1. **Observing** Which area of the onion root tip (X or Y) had the greatest number of cells undergoing mitosis? The fewest? Use specific totals from your data table to support your answer.
   X, Y; student totals will vary to support the conclusions.
2. **Predicting** If mitosis is associated with rapid growth, where do you believe is the location of most rapid root growth, area X or Y? Explain your answer.
   X; this was the area showing cells undergoing mitosis at the highest rate.
3. **Applying** Where might you look for cells in the human body that are undergoing mitosis?
   an area of rapid growth such as skin, hair follicles, intestine lining
4. **Calculating** According to your data, which phase of mitosis is most common? Least common?
   prophase (150); anaphase (2). The appearance of few cells in metaphase, anaphase, and telophase may be the result of the speed at which these phases occur.
5. **Critical Thinking** Assume that you were not able to observe cells in every phase of mitosis. Explain why this might be.
   Answers may vary—the phase has already occurred, the phase has not yet occurred, area of view is not rapidly growing, incorrect observation or recording stage.

Name Date Class

MiniLab 9-1 Use Isotopes to Understand Photosynthesis

Formulating Models

C.B. van Niel discovered the steps of photosynthesis when he used radioactive isotopes of oxygen as tracers. Radioactive isotopes are used to follow a particular molecule through a chemical reaction.

Procedure

1. Study the following equation for photosynthesis that resulted from the van Niel experiment:

$$6CO_2 + 6H_2O^* \rightarrow C_6H_{12}O_6 + 6O_2^*$$

2. Radioactive water, water tagged with an isotope of oxygen as a tracer (shown with the *), was used. Note where the tagged oxygen in water ends up on the right side of the chemical reaction.
3. Assume that van Niel repeated his experiment, but this time he put a radioactive tag on the oxygen in CO_2.
4. Using materials provided by your teacher, model what you would predict the appearance of his results would be. Your model must include a "tag" to indicate the oxygen isotope on the left side of the arrow as well as where it ends up on the right side of the arrow.
5. You must use labels or different colors in your model to indicate also the fate of carbon and hydrogen.

Analysis

1. Explain how an isotope can be used as a tag.
   Student answers may vary. Isotopes are radioactive and thus can be traced by their radioactivity.
2. Using your model, predict:
   a. the fate of all oxygen molecules that originated from carbon dioxide.
   incorporated into glucose or water
   b. the fate of all carbon molecules that originated from carbon dioxide.
   incorporated into glucose
   c. the fate of all hydrogen molecules that originated from water.
   incorporated into glucose and water

Name Date Class

MiniLab 9-2 Determine if Apple Juice Ferments

Predicting

Organisms such as yeast have the ability to break down food molecules and synthesize ATP when no oxygen is available. When the appropriate food is available, yeast can carry out alcoholic fermentation, producing CO_2. Thus, the production of CO_2 can be used to judge whether alcoholic fermentation is taking place.

Procedure

1. Carefully study the diagram on page 242 of your text and set up the experiment as shown.
2. Hold the test tube in a beaker of warm (not hot) water and observe.

Analysis

1. What were the gas bubbles that came from the plastic pipette?
   CO_2
2. Predict what would happen to the rate of bubbles given off if more yeast were present in the mixture.
   The rate would increase.
3. Why was the test tube placed in warm water?
   Warm water increases the metabolic rate of the yeast.
4. On the basis of your observations, was this process aerobic or anaerobic?
   Anaerobic, because no oxygen is available in the bulb of the pipette covered with water.

Internet BioLab

What factors influence photosynthesis?

Chapter 9

Preparation

Problem

How do different wavelengths of light a plant receives affect its rate of photosynthesis?

Objectives

In this BioLab, you will:

- **Observe** photosynthesis in an aquatic organism.
- **Measure** the rate of photosynthesis.
- **Observe** how various wavelengths of light influence the rate of photosynthesis.
- **Use the Internet** to collect and compare data from other students.

Materials

1000-mL beaker
three *Elodea* plants
string
washers
colored cellophane, assorted colors
lamp with reflector and 150-watt bulb
0.25 percent sodium hydrogen carbonate (baking soda) solution
watch with second hand

Safety Precautions

Always wear goggles in the lab.

Skill Handbook

Use the **Skill Handbook** if you need additional help with this lab.

Procedure

1. Construct a basic setup like the one shown on page 244 of your text.
2. Use the data table below to record your measurements. Be sure to include a column for each color of light you will investigate and a column for the control experiment.
3. Place the *Elodea* plants in the beaker, then completely cover the plants with water. Add some of the baking soda solution. The solution provides CO_2 for the aquarium plants. **Be sure to use the same amount of water and solution for each trial.**
4. Conduct a control experiment by directing the lamp (without colored cellophane) on the plant and notice when you see the bubbles.
5. Observe and record the number of oxygen bubbles that *Elodea* generates in five minutes.
6. Repeat steps 4 and 5 with a piece of colored cellophane. Record your observations.
7. Repeat steps 4 and 5 with a different color of cellophane and record your observations.
8. Go to the Glencoe Science Web Site at **www.glencoe.com/sec/science** to **post your data.**

Data Table

	Control	Color 1	Color 2
Bubbles observed in five minutes	**Results will vary, but will yield the highest number.**	**Results will vary depending on color of cellophane.**	**Results will vary depending on color of cellophane.**

Internet BioLab

What factors influence photosynthesis?, *continued*

Chapter 9

Analyze and Conclude

1. **Interpreting Observations** From where did the bubbles of oxygen emerge? Why?
   The bubbles emerged from the end of the *Elodea* plant.
2. **Making Inferences** Explain how counting bubbles measures the rate of photosynthesis.
   Oxygen is an end product of photosynthesis. As the rate of photosynthesis changes, so will the rate at which oxygen is produced.
3. **Using the Internet** Make a graph of your data and data posted by other students with the rate of photosynthesis per minute plotted against the wavelength of light you tested for both the control and experimental setups. Write a sentence or two explaining the graph.
   The rate for the control setup should be the greatest and therefore the highest line. The rates corresponding to colored, filtered light will be lower than the control. Results will vary depending on the color of the cellophane the students use.

Name Date Class

MiniLab 10-1 Looking at Pollen

Observing and Inferring

Pollen grains are formed within the male anthers of flowers. What is their role? Pollen contains the male gametes or sperm cells needed for fertilization. This means that pollen grains carry the hereditary units from male parent plants to female parent plants. The pollen grains that Mendel transferred from the anther of one pea plant to the pistil of another plant carried the hereditary traits that he so carefully observed in the next generation.

Procedure

1. Examine a flower. Using the diagram on page 260 of your text as a guide, locate the stamens of your flower. There are usually several stamens in each flower.
2. Remove one stamen and locate the enlarged end—the anther.
3. Add a drop of water to a microscope glass slide. Place the anther in the water. Add a coverslip. Using the eraser end of a pencil, tap the coverslip several times to squash the anther.
4. Observe under low power. Look for numerous small round structures. These are pollen grains.

Analysis

1. Provide an estimate of the number of pollen grains present in an anther.
   Numbers should be in the several thousands.
2. Describe the appearance of a single pollen grain.
   Student answers will vary depending on species used. Pollen grains are microscopic cells, often round in shape.
3. Explain the role of pollen grains in plant reproduction.
   Pollen grains contain the male reproductive cells needed for fertilization. They provide the chromosomes and genes of the male parent.

Expected Results: Students will observe that each anther contains thousands of small pollen grains.

Name Date Class

MiniLab 10-2

Formulating Models

Modeling Crossing Over

Crossing over occurs during meiosis and involves only the nonsister chromatids that are present during tetrad formation. The process is responsible for the appearance of new combinations of alleles in gamete cells.

Procedure

1. Use the data table below.
2. Roll out four long strands of clay at least 10 cm long to represent two chromosomes, each with two chromatids.
3. Use the figure on page 274 of your text as a guide in joining and labeling these model chromatids. Although there are four chromatids, assume that they started out as a single pair of homologous chromosomes prior to replication. The figure shows tetrad formation during prophase I of meiosis.
4. First, assume that no crossing over takes place. Model the appearance of the four gamete cells that will result at the end of meiosis. Record your model's appearance by drawing the gametes' chromosomes and their genes in your data table.
5. Next, repeat steps 2–4. This time, however, assume that crossing over occurs between genes B and C.

Data Table

No crossing over	Crossing Over
Appearance of gamete cells	Appearance of gamete cells

All gametes will show the same combination of genes if no crossing over occurs and different combinations of genes if crossing over does occur.

MiniLab 10-2

Modeling Crossing Over, *continued*

Analysis

1. Predict and diagram the appearance of the chromosomes prior to replication.
   Only two chromosomes should be drawn. The sequence of genes would be *C B A* on one chromosome and *c B a* on the other.
2. Define crossing over and explain when it occurs.
   Crossing over is the exchange of genetic material between nonsister chromatids during prophase I of meiosis.
3. Compare any differences in the appearance of genes on chromosomes in gamete cells when crossing over occurs and when it does not occur.
   All gamete cells show the same pattern of genes on chromosomes as in the original diagram if no crossing over occurs. If a crossing over occurs between *B* and *C*, gamete cells with gene arrangements *C B a* and *c B A* will be formed as well as cells with *C B A* and *c B a*.
4. Crossing over has been compared to "shuffling the deck" in cards. Explain what this means.
   Student answers will vary. There is a mixing of the gene traits from their original order.
5. What would be accomplished if crossing over occurred between sister chromatids? Explain your answer.
   There would be no mixing of gene traits when compared with the original chromosomes because their sister chromatids are identical.

Name Date Class

INTERNET BioLab

How can phenotypes and genotypes of plants be determined?

Chapter 10

Preparation

Problem
Can the phenotypes and genotypes of the parent plants that produced two groups of seeds be determined from the phenotypes of the plants grown from the seeds?

Hypotheses
Have your group agree on a hypothesis to be tested that will answer the problem question. Record your hypothesis.

Objective
In this BioLab, you will:
- **Analyze** the results of growing two groups of seeds.
- **Draw conclusions** about phenotypes and genotypes based on those results.
- **Use the Internet** to compare data from other schools.

Possible Materials
potting soil
small flowerpots or seedling flats
two groups of tobacco seeds
hand lens
light source
thermometer
plant-watering bottle

Safety Precautions
CAUTION: *Always wash your hands after handling plant materials.*

Skill Handbook
Use the **Skill Handbook** if you need additional help with this lab.

Plan the Experiment

1. Examine the materials provided by your teacher. As a group, make a list of the possible ways you might test your hypothesis.
2. Agree on one way that your group could investigate your hypothesis.
3. Design an experiment that will allow you to collect quantitative data. For example, how many plants do you think you will need to examine?
4. Prepare a numbered list of directions. Include a list of materials and the quantities you will need.
5. Make a data table for recording your observations.

Check the Plan.
1. Carefully determine what data you are going to collect. How many seeds do you think you will need? How long will you carry out the experiment?
2. What variables, if any, will have to be controlled? (Hint: Think about the growing conditions for the plants.)
3. ***Make sure your teacher has approved your experimental plan before you proceed.***
4. Carry out your experiment. Make any needed observations, such as the numbers of green and albino plants in each group, and complete your data table.
5. **Post your data** on the Glencoe Science Web Site at:
www.glencoe.com/sec/science

Name Date Class

INTERNET BioLab

How can phenotypes and genotypes of plants be determined?, *continued*

Chapter 10

Analyze and Conclude

1. **Thinking Critically** Why was is necessary to grow plants from the seeds in order to determine the phenotypes of the plants that formed the seeds?
Leaf color cannot be observed in the seed but appears only after the plant has emerged from the seed.
2. **Drawing Conclusions** Using the information in the introduction on page 280 of your text, describe how the gene for green color (*C*) is inherited.
The gene for green color is a dominant trait.
3. **Making Inferences** For the group of seeds that yielded all green plants, are you able to determine exactly the genotypes of the parents that formed these seeds? Can you determine the genotype of each plant observed? Explain.
No, one parent may have been true breeding for green (*CC*), the other may have been heterozygous (*Cc*). This would have yielded all green offspring. Offspring may be *CC* or *Cc* but still appear green.
4. **Making Inferences** For the group of seeds that yielded some green and some albino plants, are you able to determine exactly the genotypes of the plants that formed these seeds? Can you determine the genotype of each plant observed? Explain.
Yes, both parents must be heterozygous to yield a ratio of 3 green to 1 albino. For the offspring genotypes, you can conclude only that the albino offspring are *cc*. Green are either *CC* or *Cc*.
5. **Using the Internet** Compare your experimental design with that of other students. Were your results similar? What might account for the differences?
Answers may vary. Genetic ratios are governed by the laws of probability. The larger the population size, the closer the calculated value will be to the theoretical.

Data and Observations: Seeds that came from true breeding plants will produce plants that are all green. Seeds from heterozygous parents will produce both green and albino seedlings in the ratio of about 3 green to 1 albino. Have students review Problem-Solving Lab 10-1 for help in calculating the ratio of green to albino plants.

Name Date Class

MiniLab 11-1 Transcribe and Translate

Predicting

Molecules of DNA carry the genetic instructions for protein formation. Converting these DNA instructions into proteins requires a series of coordinated steps in transcription and translation.

Procedure

1. Use the data table below.
2. Complete column B by writing the correct mRNA codon for each sequence of DNA bases listed in the column marked *DNA Base Sequence*. Use the letters A, U, C, or G.
3. Identify the process responsible by writing its name on the arrow in column A.
4. Complete column D by writing the correct anticodon that bonds to each codon from column B.
5. Identify the process responsible by writing its name on the arrow in column C.
6. Complete column E by writing the name of the correct amino acid that is coded by each base sequence. Use *Table 11.2* on page 298 of your text to translate the mRNA base sequences to amino acids.

Data Table

	A	B	C	D	E
DNA base sequence	Process	mRNA codon	Process	tRNA anticodon	Amino acid
AAT	**transcription** →	**UUA**	**translation** →	**AAU**	**leucine**
GGG	**transcription** →	**CCC**	**translation** →	**GGG**	**proline**
ATA	**transcription** →	**UAU**	**translation** →	**AUA**	**tryosine**
AAA	**transcription** →	**UUU**	**translation** →	**AAA**	**phenylalanine**
GTT	**transcription** →	**CAA**	**translation** →	**GUU**	**glutamine**

Name Date Class

MiniLab 11-1 Transcribe and Translate, *continued*

Analysis

1. Where within the cell:
   a. are the DNA instructions located?
   on chromosomes in the nucleus
   b. does transcription occur?
   in the nucleus
   c. does translation occur?
   in the ribosomes
2. Describe the structure of a tRNA molecule.
   tRNA is a small molecule that has a three-base anticodon at one end and an amino-acid attachment site at the opposite end.
3. Explain why specific base pairing is essential to the processes of transcription and translation.
   Precise base pairing is essential to transcription and translation so that the correct genetic information in DNA is transferred to the forming protein.

Name Date Class

MiniLab 11-2

Making and Using Tables

Gene Mutations and Proteins

Gene mutations often have serious effects on proteins. In this activity, you will demonstrate how such mutations affect protein synthesis.

Procedure

1. Use the following base sequence of one strand of an imaginary DNA molecule: AATGCCAGTGGTTCGCAC.
2. Write the base sequence for an mRNA strand that would be transcribed from the given DNA sequence.
3. Use *Table 11.2* to determine the sequence of amino acids in the resulting protein fragment.
4. If the fourth base in the original DNA strand were changed from G to C, how would this affect the resulting protein fragment?
5. If a G were added to the original DNA strand after the third base, what would the resulting mRNA look like? How would this addition affect the protein?

Analysis

1. Which change in DNA was a point mutation? Which was a frameshift mutation?

   Changing G to C was a point mutation; adding G to the chain was a frameshift mutation.

2. In what way did the point mutation affect the protein?

   The point mutation changed only one amino acid.

3. How did the frameshift mutation affect the protein?

   The frameshift mutation changed every amino acid following the addition of G.

Expected Results:

Base sequence of mRNA:

UUACGGUCACCAAGCGUG; amino acids; leucine-arginine-serine-proline-serine-valine; Changing the fourth base in the DNA from G to C would change the second amino acid from arginine to glycine. Adding G at the fourth position of the DNA would result in the mRNA base sequence UUACCGGUCACCAAGCGUG and the amino acid sequence leucine-proline-valine-threonine-lysine-arginine.

Name Date Class

INVESTIGATE BioLab

RNA Transcription

Chapter 11

Preparation

Problem

How does the order of bases in DNA determine the order of bases in mRNA?

Objectives

In this BioLab, you will:

- **Formulate a model** to show how the order of bases in DNA determines the order of bases in mRNA.
- **Infer** why the structure of DNA enables it to be easily copied.

Materials

construction paper, 5 colors
scissors
clear tape
pencil

Safety Precautions

Be careful when using scissors. Always wear goggles in the lab.

Skill Handbook

Use the **Skill Handbook** if you need additional help with this lab.

Procedure

1. Copy the illustrations of the four different DNA nucleotides on page 308 of your text onto your construction paper, making sure that each different nucleotide is on a different color paper. You should make ten copies of each nucleotide.
2. Using scissors, carefully cut out the shapes of each nucleotide.
3. Using any order of nucleotides that you wish, construct a double-stranded DNA molecule. If you need more nucleotides, copy them as before.
4. Fasten your molecule together using clear tape. Do not tape across base pairs.
5. As in step 1, copy the illustrations of A, G, and C nucleotides. Use the same colors of construction paper as in step 1. Use the fifth color of construction paper to make copies of uracil nucleotides.
6. With scissors, carefully cut out the nucleotide shapes.
7. With your DNA molecule in front to you, demonstrate the process of transcription by first pulling the DNA molecule apart between the base pairs.
8. Using only one of the strands of DNA, begin matching complementary mRNA nucleotides with the exposed bases on the DNA model to make RNA.
9. When you are finished, tape your new mRNA molecule together.

Name Date Class

INVESTIGATE BioLab

RNA Transcription, *continued*

Chapter 11

Analyze and Conclude

1. **Observing and Inferring** Does the mRNA model more closely resemble the DNA strand from which it was transcribed or the complementary strand that wasn't used? Explain your answer.
   mRNA more closely resembles the complementary DNA. They have the same base sequence except that mRNA has a uracil in place of thymine.
2. **Recognizing Cause and Effect** Explain how the structure of DNA enables the molecule to be easily transcribed. Why is this important for genetic information?
   Because DNA is double-stranded, sections can be unzipped to allow complementary bases to hydrogen bond while the remaining DNA stays zipped. Thus, only the information needed at one time is being transcribed.
3. **Relating Concepts** Why is RNA important to the cell? How does an mRNA molecule carry information from DNA?
   The mRNA is formed as a complementary copy of the genetic information. The RNA copy can leave the nucleus while the "master copy" stays protected within the nucleus.

Name Date Class

MiniLab 12-1 Illustrating a Pedigree

Analyzing Information

The pedigree method of studying a trait in a family uses records of phenotypes extending for two or more generations. Studies of pedigrees can be used to yield a great deal of genetic information about a related group.

Procedure

1. Working with a partner, choose one human trait, such as attached and free-hanging earlobes or tongue rolling, that interests both of you.
2. Using either your or your partner's family, collect information about your chosen trait. Include whether each individual is male or female, does or does not have the trait, and the relationship of the individual to others in the family.
3. Use your information to draw a pedigree for the trait.
4. Try to determine how your trait is inherited.

Analysis

1. What trait did you study? Can you determine from your pedigree what the apparent inheritance pattern of the trait is?
   Answers may include earlobe shape, widow's peak, tongue rolling, or ability to taste PTC paper.
2. How is the study of inheritance patterns limited by pedigree analysis?
   The number of individuals may be too small to determine the inheritance pattern.

Name Date Class

MiniLab 12-2 Detecting Colors and Patterns in Eyes

Observing and Inferring

Human eye color, like skin color, is determined by polygenic inheritance. You can detect several shades of eye color, especially if you look closely at the iris with a magnifying glass. Often, the pigment is deposited so that light reflects from the eye, causing the iris to appear blue, green, gray, or hazel (brown-green). In actuality, the pigment may be yellowish or brown, but not blue.

Procedure

1. Use a magnifying glass to observe the patterns and colors of pigment in the eyes of five classmates.
2. Use colored pencils to make drawings of the five irises.
3. Describe your observations in your journal.

Analysis

1. How many different pigments were you able to detect in each eye?
   Students may detect brown, blue, yellow, gray, green, and black pigments in each eye.
2. From your data, do you suspect that eye color might not be inherited by simple Mendelian rules? Explain.
   Eye color is probably not inherited by simple Mendelian rules because there are so many phenotypes.
3. Suppose that two people have brown eyes. They have two children with brown eyes, one with blue eyes, and one with green eyes. What pattern might this suggest?
   The pattern suggests polygenic inheritance.

Expected Results: Students will see Barr bodies in cells from females but not in those from males.

Name Date Class

DESIGN YOUR OWN BioLab

What is the pattern of cytoplasmic inheritance?

Chapter 12

Preparation

Problem
What inheritance pattern does the variegated leaf trait in *Brassica* show?

Hypotheses
Consider the possible evidence you could collect that would answer the problem question. Among the people in your group, form a hypothesis that you can test to answer the question, and write the hypothesis in your journal.

Objectives
In this BioLab, you will:
- **Determine** which crosses of *Brassica* plants will reveal the pattern of cytoplasmic inheritance.
- **Analyze** data from *Brassica* crosses.

Possible Materials
Brassica rapa seeds, normal and variegated
potting soil and trays
paintbrushes
forceps
single-edge razor blade
light source
labels

Safety Precautions
Always wear goggles in the lab. Handle the razor blade with extreme caution. Always cut away from you. Wash your hands with soap and water after working with plant material.

Skill Handbook
Use the **Skill Handbook** if you need additional help with this lab.

Plan the Experiment

1. Decide which crosses will be needed to test your hypothesis.
2. Keep the available materials in mind as you plan your procedure. How many seeds will you need?
3. Record your procedure, and list the materials and quantities you will need.
4. Assign a task to each member of the group. One person should write data in a journal, another can pollinate the flowers, while a third can set up the plant trays. Determine who will set up and clean up materials.
5. Design and construct a data table for recording your observations.

Check the Plan
Discuss the following points with other group members to decide the final procedure for your experiment.

1. What data will you collect, and how will data be recorded?
2. When will you pollinate the flowers? How many flowers will you pollinate?
3. How will you transfer pollen from one flower to another?
4. How and when will you collect the seeds that result from your crosses?
5. What variables will have to be controlled? What controls will be used?
6. When will you end the experiment?
7. ***Make sure your teacher has approved your experimental plan before you proceed further.***
8. Carry out your experiment.

Name Date Class

DESIGN YOUR OWN BioLab

What is the pattern of cytoplasmic inheritance?, *continued*

Chapter 12

Analyze and Conclude

1. **Checking Your Hypothesis** Did your data support your hypothesis? Why or why not?
   Student answers will vary depending on their hypotheses.
2. **Interpreting Observations** What is the inheritance pattern of variegated leaves in *Brassica?*
   Variegation is inherited as a cytoplasmic trait from the female parent.
3. **Making Inferences** Explain why genes in the chloroplast are inherited in this pattern.
   Sperm cells contain little to no cytoplasm. Cytoplasm containing chloroplasts is contributed only by the egg.
4. **Drawing Conclusions** Which parent is responsible for passing the variegated trait to its offspring?
   female
5. **Making Scientific Illustrations** Draw a diagram tracing the inheritance of this trait through cell division.
   Diagrams should show the trait being transmitted in an egg cell of the female but not in pollen of the male.

Data and Observations

Variegated F_1 plants will appear in crosses where the female parent was variegated.

Name Date Class

MiniLab 13-1 Matching Restriction Enzymes to Cleavage Sites

Applying Concepts

Many restriction enzymes cut sequences of DNA that are palindromes. As a result of cuts to the DNA, single-stranded sequences of DNA are left dangling at the ends of a fragment. These ends are available for pairing with their complementary bases in a plasmid or piece of viral DNA.

Procedure

1. Use the data table below.

Data Table

DNA fragment	Enzyme letter (D–F)	Action of restriction enzyme	Cleaved fragment of DNA
—GGTACC— / —CCATGG—	E	—GGTACC— / —CCATGG—	—G GTACC— / —CCATG G—
—CCATGG— / —GGTACC—	E	—CCATGG— / —GGTACC—	—CCATG G— / —G GTACC—
—CAATTG— / —GTTAAC—	D	—CAATTG— / —GTTAAC—	—CA ATTG— / —GTTA AC—
—GATATC— / —CTATAG—	F	—GATATC— / —CTATAG—	—G ATATC— / —CTATA G—

2. Figure out which restriction enzyme will cleave each DNA fragment. Use the following guides.
   Enzyme D cleaves at an A-A site and leaves 3 single-stranded bases on each end.
   Enzyme E cleaves at a G-G site and leaves 4 single-stranded bases on each end.
   Enzyme F cleaves at a G-A site and leaves 4 single-stranded bases on each end.
3. Draw in the action of each enzyme. Record its letter.
4. Diagram each fragment of DNA as it would appear if cleaved by the proper restriction enzyme.
5. Use the top row in the table as an example and guide.

Name Date Class

MiniLab 13-1 Matching Restriction Enzymes to Cleavage Sites, *continued*

Analysis

1. Use the example provided in the data table to illustrate a single-stranded dangling end of DNA.
   GTAC– is a dangling end.
2. Record the DNA base sequence that must be present on a piece of viral DNA if these ends could "stick to" the dangling bases in the example shown in the data table.
   –CATG
3. Are restriction enzymes very specific as to where they cleave DNA? Explain your answer and give an example.
   Yes; each enzyme can cleave only a specific site and can leave only a specific number of single-stranded DNA bases. Enzyme F cleaves between G and A but must leave four bases in the dangling end.

Name Date Class

MiniLab 13-2

Storing the Human Genome

Using Numbers

It has been estimated that the human genome consists of three billion nitrogen base pairs. How much room would all the genetic information in a single cell take up if it were printed in a book the size of a typical novel?

Procedure

1. Use the data table below.
2. Select a random page from a novel.
3. Follow the directions in the table. Record your calculations in your data table.

Data Table

Directions	Letters and numbers
A. Count the number of characters (letters, punctuation marks, and spaces) across one entire line of your selected page.	**Sample data 80**
B. Count the number of lines on the page.	**45**
C. Calculate the number of characters on the page. (Multiply A × B.)	**3600**
D. Let one nitrogen base equal one character. Knowing that DNA is made of nitrogen base pairs, divide C by 2.	**1800**
E. Record the number of pages in your novel.	**400**
F. Calculate the number of base pairs in your novel. (Multiply E × D.)	**720 000**
G. Calculate the number of books the size of your novel needed to hold the human genome. (Divide 3 billion by F.)	**4167**

Analysis

1. What changes could be taken to improve the accuracy of this activity at steps A–C?
Take an average of several pages to arrive at total number of characters per page.

2. What assumption is being made at step G?
The human genome is made up of 3 billion bases.

Name Date Class

MiniLab 13-2

Storing the Human Genome, *continued*

3. Explain the logic for step D.
DNA bases are paired (A-T, C-G). Thus, the total genome is expressed in pairs so the number of characters must be expressed in pairs.

4. a. How many books the size of your novel would be needed to store the human genome?
4167 books

b. How many books the size of your novel would be needed to store a typical bacterial genome? Assume there are three million base pairs in the genome of a bacterium.
4.17 books

Name Date Class

Investigate BioLab

Modeling Recombinant DNA

Chapter 13

Preparation

Problem
How can you model recombinant DNA technology?

Objectives
In this BioLab, you will:
- **Model** the process of preparing recombinant DNA.
- **Analyze** a model for preparation of recombinant DNA.

Materials
white paper
colored pencils, red and green
tape
scissors

Safety Precautions
Always wear goggles in the lab. Be careful with sharp objects.

Skill Handbook
Use the **Skill Handbook** if you need additional help with this lab.

Procedure

1. Cut a lengthwise strip of paper from a sheet of white paper into a rectangle about 3 cm by 28 cm. This strip represents a long sequence of DNA containing a particular gene that you wish to combine with a plasmid.
2. Cut another lengthwise strip of paper into a rectangle about 3 cm by 10 cm. When taped into a ring, this piece of paper will represent a bacterial plasmid.
3. Use your colored pencils to color the longer strip red and the shorter strip green.
4. Write the following DNA sequence once on the shorter strip of paper and two times about 5 cm apart on the longer strip of paper.

   -G-G-A-T-C-C-
   -C-C-T-A-G-G-
5. After coloring the shorter strip of paper and writing the sequence on it, tape the ends together.
6. Assume that a particular restriction enzyme is able to cleave DNA in a staggered way as illustrated here.

   -G   G-A-T-C-C-
   -C-C-T-A-G   G-

   Cut the longer strand of DNA in both places. You now have a cleaved foreign DNA fragment containing a gene that can be inserted into the plasmid.
7. Once the sequence containing the foreign gene has been cleaved, cut the plasmid in the same way.
8. Splice the foreign gene into the plasmid by taping the paper together where the sticky ends pair properly. The new plasmid represents recombinant DNA.

Name Date Class

Investigate BioLab

Modeling Recombinant DNA, *continued*

Chapter 13

9. Use the data table at right. Relate the steps of producing recombinant DNA to the activities of the modeling procedure by explaining how the terms relate to the model.

Data Table

Term	BioLab model
Gene splicing	process of taping green and red paper together
Plasmid	green strip
Restriction enzyme	scissors
Sticky ends	cut ends on paper
Recombinant DNA	red and green strips taped together

Analyze and Conclude

1. **Comparing and Contrasting** How does the paper model of a plasmid resemble a bacterial plasmid?
   It is a small circular piece of DNA.
2. **Comparing and Contrasting** How is cutting with the scissors different from cleaving with a restriction enzyme?
   The scissors can cut the DNA anywhere, but the restriction enzyme recognizes a particular sequence.
3. **Thinking Critically** Enzymes that modify DNA, such as restriction enzymes, have been discovered and isolated from living cells. What functions do you think they have in living cells?
   These enzymes may function in a cell by cutting up and destroying invading viral DNA.

Name Date Class

MiniLab 14-1 Marine Fossils

Observing and Inferring

Certain sedimentary rocks are formed totally from the fossils of once-living ocean organisms called diatoms. The diatom fossils are often 1000 meters thick. These sedimentary rocks were at one time in the past under ocean water and were then lifted above sea level during periods of geological change.

Procedure

1. Prepare a wet mount of a small amount of diatomaceous earth. **CAUTION:** ***Use care in handling microscope slides and coverslips.***
2. Examine the material under low-power magnification.
3. Draw several of the different shapes you see.
4. Compare the shapes of the fossils you observe to present-day diatoms shown in the photograph. Remember, however, that the fossils you observe are probably only pieces of the whole organism.

Analysis

1. Describe the appearance of fossil diatoms.
   Answers will vary—rod-shaped, glasslike, circular, boatlike, needle-shaped, ridged or scored surface
2. How are fossil diatoms similar to and different from the diatoms in the photo? Can you use these similarities and differences to predict how diatoms have changed over time? Explain your answer.
   Although broken, the fossil diatoms look similar and therefore have probably changed little over time.
3. What part of the original diatom did you observe under the microscope? How did this part survive millions of years? Why were the fossils you observed in pieces?
   The cell walls were visible because they did not decompose. The weight of water and sediments crushed them.

Expected Results: Students should see broken cell walls of many different shapes.

Name Date Class

MiniLab 14-2 A Time Line

Organizing Data

In this activity, you will construct a time line that is a scale model of the Geologic Time Scale. Use a scale in which 1 meter equals 1 billion years. Each millimeter then represents 1 million years.

Procedure

1. Use a meterstick to draw a continuous line down the middle of a 5 m strip of adding-machine tape.
2. At one end of the tape, draw a vertical line and label it "The Present."
3. Measure off the distance that represents 4.6 billion years ago. Draw a vertical line at that point and label it "Earth's Beginning."
4. Using the table below, plot the location of each event on your time line. Label the event and the number of years ago it occurred.

Geologic Time Scale

Event	Estimated years ago	Event	Estimated years ago
Earliest evidence of life	3.5 billion	First birds	150 million
Paleozoic era begins	540 million	Cretaceous period begins	146 million
First land plants	430 million	Dinosaurs become extinct	66 million
Mesozoic era begins	245 million	Cenozoic era begins	66 million
Triassic period begins	245 million	Primates appear	60 million
Jurassic period begins	208 million	Humans appear	200 000
First dinosaurs	225 million		

Analysis

1. Which era is the longest? The shortest?
   the longest era—Paleozoic; the shortest era—Cenozoic
2. In which eras did dinosaurs and mammals appear on Earth?
   Mesozoic era
3. Which lived on Earth the longer time, dinosaurs or mammals?
   mammals

Name Date Class

INVESTIGATE BioLab

Determining a Fossil's Age

Chapter 14

Preparation

Problem
How can you simulate radioactive half-life?

Objectives
In this BioLab you will:
- **Formulate models** Simulate the radioactive decay of K 40 into Ar 40 with pennies.
- **Collect data** Collect data to determine the amount of K 40 present after several half-lives.
- **Make and use a graph** Graph your data and use its values to determine the age of rocks.

Materials
shoe box with lid
100 pennies
graph paper

Skill Handbook
Use the **Skill Handbook** if you need additional help with this lab.

Procedure

1. Use the data table below.
2. Place 100 pennies in a shoe box.
3. Arrange the pennies so that their "head" sides are facing up. Each "head" represents an atom of K 40, and each "tail" an atom of Ar 40.
4. Record the number of "heads" and "tails" present at the start of the experiment. Use the row marked "0" in the data table.
5. Cover the box. Then shake the box well. Let the shake represent one half-life of K 40, which is 1.3 billion years.

Data Table

Number of Shakes (half lives)	Number of Heads (K 40 atoms left)				
	Trial 1	Trial 2	Trial 3	Totals	Average
0					100
1					50
2			Data may	vary.	25
3					12
4					6
5					3

6. Remove the lid and record the number of "heads" you see facing up. Remove all the "tail" pennies.
7. To complete the first trial, repeat steps 5 and 6 four more times.
8. Run two more trials and determine an average for the number of "heads" present at each half-life.
9. Draw a full-page graph. Plot your average values on the graph. Plot the number of half-lives for K 40 on the x-axis and the number of "heads" on the y-axis. Connect the points with a line. Remember, each half-life mark on the graph axis for K 40 represents 1.3 billion years.

Name Date Class

INVESTIGATE BioLab

Determining a Fossil's Age, *continued*

Chapter 14

Analyze and Conclude

1. **Applying Concepts** What symbol represented an atom of K 40 in this experiment? What symbol represented an atom of Ar 40?
K-40—penny with "head" side up; Ar-40—penny with "tail" side up

2. **Thinking Critically** Compare the numbers of protons and neutrons of K 40 and Ar 40. (Consult the Periodic Chart in the Appendix for help.) Can Ar 40 change back to K 40? Explain your answer, pointing out what procedural part of the experiment supports your answer.
Potassium has 19 protons and 21 neutrons, and Argon has 18 protons and 22 neutrons; No. K40 decays or changes into Ar-40. Remove "tail" pennies.

3. **Defining Operationally** Define the term half-life. What procedural part of the simulation represented a half-life period of time in the experiment?
A half-life is the time needed for half the number of atoms of a radioactive element to change into atoms of a different element. Shaking the box represented a half-life.

4. **Communicating** Explain how scientists use radioactive dating to approximate a fossil's age.
They compare the amount of a radioactive element present now in a fossil or rock to the amount originally present, and use the element's half-life to calculate the sample's age.

5. **Making and Using Graphs** You are attempting to determine the age of a rock sample. Use your graph to read the rock's age if it has:
a. 70% of its original K 40 amount.
650 000 000
b. 35% of its original K 40 amount.
1 950 000 000
c. 10% of its original K 40 amount.
4 550 000 000

Name Date Class

MiniLab 15-1

Formulating Models

Camouflage Provides an Adaptive Advantage

Camouflage is a structural adaptation that allows organisms to blend with their surroundings. In this activity, you'll discover how natural selection can result in camouflage adaptations in organisms.

Procedure

1. Working with a partner, punch 100 dots from a sheet of white paper with a paper hole punch. Repeat with a sheet of black paper. These dots will represent black and white insects.
2. Scatter both white and black dots on a sheet of black paper.
3. Decide whether you or your partner will role-play a bird.
4. The "bird" looks away from the paper, then turns back, and immediately picks up the first dot he or she sees.
5. Repeat step 4 for one minute.

Analysis

1. What color dots were most often collected?

white dots

2. How does color affect the survival rate of insects?

Light-colored insects may be seen and preyed on more easily than dark-colored insects. Therefore, dark-colored insects have a higher survival rate.

3. What might happen over many generations to a similar population in nature?

Over time, an insect population might become dark-colored because light-colored insects were eliminated from the population.

Expected Results: Most groups will have picked up more white dots than black dots.

Name Date Class

MiniLab 15-2

Collecting Data

Detecting a Variation

Pick almost any trait—height, eye color, leaf width, or seed size—and you can observe how the trait varies in a population. Some variations are an advantage to an organism and some are not.

Procedure

1. Copy the data table shown here, but include the lengths in millimeters (numbers 25 through 45) that are missing from this table.

Data Table

Length in mm	20	21	22	23	24	–	46	47	48	49	50
Checks	**Student data should indicate a wide range**										
My Data—Number of shells	**in shell length. There should be few shells at**										
Class Data—Number of shells	**either extreme and the majority of shells should fall in the middle range of the measurements.**										

2. Use a millimeter ruler to measure a peanut shell's length. In the Checks row, check the length you measured.
3. Repeat step 2 for 29 more shells.
4. Count the checks under the length and enter the total in the row marked My Data.
5. Use class totals to complete the row marked Class Data.

Analysis

1. Was there variation among the lengths of peanut shells? Use specific class totals to support your answer.

yes—student answers will vary

2. If larger peanut shells were a selective advantage, would this be stabilizing, directional, or disruptive selection? Explain your answer.

Directional selection—larger shells may favor the survival of offspring because they may contain larger, more viable seeds.

Name Date Class

INTERNET BioLab

Natural Selection and Allelic Frequency

Chapter 15

Preparation

Problem
How does natural selection affect allelic frequency?

Objectives
In this BioLab, you will:
- **Simulate** natural selection by using beans of two different colors.
- **Calculate** allelic frequencies over five generations.
- **Demonstrate** how natural selection can affect allelic frequencies over time.
- **Use the Internet** to collect and compare data from other students.

Materials
colored pencils (2)
paper bag
graph paper
pinto beans
white navy beans

Skill Handbook
Use the **Skill Handbook** if you need additional help with this lab.

Procedure

1. Use the data table.
2. Place 50 pinto beans and 50 white navy beans into the paper bag.
3. Shake the bag. Remove two beans. These represent one rabbit's genotype. Set the pair aside, and continue to remove 49 more pairs.
4. Arrange the beans on a flat surface in two columns representing the two possible rabbit phenotypes, gray (genotypes *GG* or *Gg*) and white (genotype *gg*).
5. Examine your columns. Remove 25 percent of the gray rabbits and 100 percent of the white rabbits. These numbers represent a random selection pressure on your rabbit population. If the number you calculate is a fraction, remove a whole rabbit to make whole numbers.
6. Count the number of pinto and navy beans remaining. Record this number in your data table.
7. Calculate the allelic frequencies by dividing the number of beans of one type by 100. Record these numbers in your data table.
8. Begin the next generation by placing 100 beans into the bag. The proportions of pinto and navy beans should be the same as the percentages you calculated in step 7.
9. Repeat steps 3 through 8, collecting data for five generations.
10. **Post your data** on the Glencoe Science Web Site at **www.glencoe.com/sec/science.**
11. Graph the frequencies of each allele over five generations. Plot the frequency of the allele on the vertical axis and the number of the generation on the horizontal axis. Use a different colored pencil for each allele.

Name Date Class

INTERNET BioLab

Natural Selection and Allelic Frequency, *continued*

Chapter 15

Data Table

	Allele *G*			Allele *g*		
Generation	Number	Percentage	Frequency	Number	Percentage	Frequency
Start	50	50	0.50	50	50	0.50
1						
2						
3						
4						
5						

Analyze and Conclude

1. **Analyzing Data** Did either allele disappear? Why or why not?
Neither allele disappeared from the population because the *g* allele is also in the heterozygous *(Gg)* rabbits.
2. **Thinking Critically** What does your graph show about allelic frequencies and natural selection?
The graph shows an increase in the frequency of the *G* allele and a decrease in the frequency of the *g* allele due to natural selection against white rabbits.
3. **Making Inferences** What would happen to the allelic frequencies if the number of eagles declined?
There would be less selective pressure on white rabbits and, therefore, less decline in the frequency of the *g* allele.
4. **Interpreting Data** Explain any differences in allelic frequencies you observed between your data and the data from the Internet.
Students should notice little difference in the allelic frequencies posted on the Internet and the frequencies they calculated. By combining data, students may get more accurate results.

Data and Observations: Make sure students are correctly calculating allelic frequency after each "generation" and recording these data in their data tables. Students should observe changes in the allelic frequencies of the rabbit population. Student graphs should show an increase in the frequency of the *G* allele and a decrease in the *g* allele.

Name Date Class

MiniLab 16-1

Comparing and Contrasting

Comparing Old and New World Monkeys

In this activity, you will gather and then compare data about Old World monkeys and New World monkeys.

Procedure

1. Use the data table below.
2. Examine the diagrams on page 433 of your text.
3. Complete the data table.

Analysis

1. Why would a low body weight be helpful for an animal with an arboreal life style?
   It would allow for efficient movement in trees.
2. Which group appears to be more closely related to humans? Explain your answer. (Note: Humans have eight premolars, twelve molars, and a total of 32 teeth.)
   Old World monkeys share more traits with humans.

Data Table

Characteristic	New World monkey	Old World monkey
Number of premolars ($\frac{1}{4}$ jaw)	**3**	**2**
Number of molars ($\frac{1}{4}$ jaw)	**3**	**3**
Total teeth in mouth	**36**	**32**
Nostril position	**up**	**down**
Tail	**prehensile**	**not prehensile**
Body weight	0.14 to 11 kg	1.2 to 30 kg

Name Date Class

MiniLab 16-2

Analyzing Information

Compare Human Proteins with Those of Other Primates

Scientists use differences in amino acid sequences in proteins to determine evolutionary relationships of living species. In this activity, you'll compare representative short sequences of amino acids of a protein among other groups of primates to determine their evolutionary history.

Procedure

1. Use the data table below.
2. For each primate listed in the table at right, determine how many amino acids differ from the human sequence. Record these numbers in the data table.
3. Calculate the percentage differences by dividing the numbers by 15 and multiplying by 100. Record the numbers in your data table.

Table 16.1 Amino Acid Sequences in Primates

Baboon	Chimp	Lemur	Gorilla	Human
ASN	SER	ALA	SER	SER
THR	THR	THR	THR	THR
THR	ALA	SER	ALA	ALA
GLY	GLY	GLY	GLY	GLY
ASP	ASP	GLU	ASP	ASP
GLU	GLU	LYS	GLU	GLU
VAL	VAL	VAL	VAL	VAL
ASP	GLU	GLU	GLU	GLU
ASP	ASP	ASP	ASP	ASP
SER	THR	SER	THR	THR
PRO	PRO	PRO	PRO	PRO
GLY	GLY	GLY	GLY	GLY
GLY	GLY	SER	GLY	GLY
ASN	ALA	HIS	ALA	ALA
ASN	ASN	ASN	ASN	ASN

Analysis

1. Which primate is most closely related to humans? Least closely related?
   gorilla and chimpanzee; lemur
2. On another sheet of paper, construct a diagram of primate evolutionary relationships that most closely fits your results.
   Baboons should branch off lemurs. Gorillas, chimpanzees, and humans should be close together.

Data Table

Primate	Amino acids different from humans	Percent difference
Baboon		
Chimpanzee		
Gorilla		
Lemur		

Expected Results: The human, chimpanzee, and gorilla sequences are identical. Baboons differ by 33 percent and lemurs by 47 percent.

Name Date Class

INVESTIGATE BioLab

Comparing Skulls of Three Primates

Chapter 16

Preparation

Problem
How do skulls of primates provide evidence for human evolution?

Objectives
In this BioLab, you will:
- **Determine** how paleoanthropologists study early human ancestors.
- **Compare and contrast** the skulls of australopithecines, gorillas, and modern humans.

Materials
metric ruler
protractor
copy of skull diagrams

Skill Handbook
Use the **Skill Handbook** if you need additional help with this lab.

Procedure

1. Your teacher will provide copies (1/2 natural size) of the skulls of *Australopithecus africanus*, *Gorilla gorilla*, and *Homo sapiens*.
2. The rectangles drawn over the skulls represent the areas of the brain (upper rectangle) and face (lower rectangle). On each skull, determine and record the area of each rectangle (length × width).
3. Measure the diameters of the circles in each skull. Multiply these numbers by 200 cm^2. The result is the cranial capacity (brain volume) in cubic centimeters.
4. The two heavy lines projected on the skulls are used to measure how far forward the jaw protrudes. Use your protractor to measure the outside angle (toward the right) formed by the two lines.
5. Complete the data table.

Data Table

	Gorilla	Australopithecus	Modern human
1. Face area in cm^2	32 cm^2	19 cm^2	12 cm^2
2. Brain area in cm^2	23 cm^2	23 cm^2	40 cm^2
3. Is brain area smaller or larger than face area?	Smaller	Larger	Larger
4. Is brain area 3 times larger than face area?	No	No	Yes
5. Cranial capacity in cm^3	600 cm^3	600 cm^3	1060 cm^3
6. Jaw angle	35	55	90
7. Does lower jaw stick out in front of nose?	Yes	Yes	No
8. Is sagittal crest present?	Yes	Yes	No
9. Is browridge present?	Yes	Yes	No

INVESTIGATE BioLab

Comparing Skulls of Three Primates, *continued*

Chapter 16

Analyze and Conclude

1. **Comparing and Contrasting** How would you describe the similarities and differences in face-to-brain area in the three primates?
Humans have a small facial area compared with brain area. Apes have a large facial area compared with brain area. Australopithecines were intermediate between apes and humans but closer to apes.

2. **Interpreting Observations** How do the cranial capacities compare among the three skulls? How do the jaw angles compare?
Apes have a small cranial capacity, whereas humans have a large cranial capacity, and that of australopithecines was intermediate but closer to the apes. An ape has a small jaw angle, and a human has a large jaw angle. The jaw angle of australopithecines was intermediate but closer to that of the ape.

3. **Drawing Conclusions** Based on your findings, what statements can you make about the placement of australopithecines in human evolution?
Many australopithecine skull traits were intermediate between those of apes and humans, and some were more similar to those of apes. Australopithecines represent very early human ancestors.

Name Date Class

MiniLab 17-1 Using a Dichotomous Key

Classifying

How could you identify a tree growing in front of your school? You might ask a local expert, or you could use a manual or field guide that contains descriptive information and keys about trees. A key is a set of descriptive sentences that is subdivided into steps. A dichotomous key has two descriptions at each step. You follow the steps until the key reveals the name of the tree.

Procedure

1. Using a few leaves from local trees and a dichotomous key for trees of your area, identify the tree from which each leaf came. To use the key, study one leaf. Then choose the one statement from the first pair that most accurately describes the leaf. Continue following the key until you identify the leaf's tree. Repeat the process for each leaf.
2. Glue each leaf on a separate sheet of paper. For each leaf, record the tree's name.

Analysis

1. What is the function of a dichotomous key?
   identification of organisms
2. List three different characteristics used in your key.
   vein and margin structure, leaf shape and size, number of lobes
3. As you used the key, did the characteristics become more general or more specific?
   more specific

Name Date Class

MiniLab 17-2 Using a Cladogram to Show Relationships

Classifying

Cladograms were developed by Willie Hennig. They use derived characteristics to illustrate evolutionary relationships.

Procedure

1. The following table shows the presence or absence of six derived traits in the seven dinosaurs that are labeled A–G.
2. Use the information listed in the table to answer the questions below.

Derived traits of dinosaurs

Dinosaur trait	A	B	C	D	E	F	G
Hole in hip socket	yes	yes	yes	yes	yes	yes	yes
Extension of pubis bone	no	no	no	yes	yes	yes	yes
Unequal enamel on teeth	no	no	no	no	yes	yes	yes
Skull has "shelf" in back	no	no	no	no	no	yes	yes
Grasping hand	yes	yes	yes	no	no	no	no
Three-toed hind foot	yes	yes	no	no	no	no	no

Expected Results: Students will complete the cladogram as follows: node 1 = hole in hip socket, node 4 = extension of pubis bone, node 5 = unequal enamel, node 6 = skull shelf.

Analysis

1. Copy the partially completed cladogram on page 467 of your text. Complete the missing information on the right side.
   See "Expected Results" for correct completed cladogram.
2. How many traits does dinosaur F share with dinosaur C, with dinosaur D, and with dinosaur E?
   Dinosaur F shares 1 trait with C, 2 traits with D, and 3 traits with E.
3. Dinosaurs A and B form a grouping called a clade. Dinosaurs A, B, and C form another clade. What derived trait is shared only by the A and B clade? By the A, B, and C clade? By the D, E, F, and G clade?
   three-toed hind foot; grasping hand; extension of pubis bone
4. Traits that evolved early, such as the hole in the hip socket, are called primitive traits. Traits that evolved later, such as a grasping hand, are called derived traits. Are primitive traits typical of broader or smaller clades? Are derived traits typical of broader or smaller clades? Give an example in each case.
   broader—hole in hip socket extends to all 7 animals; smaller—three-toed hind foot extends to only 1 clad

Name Date Class

INVESTIGATE BioLab

Making a Dichotomous Key

Chapter 17

Preparation

Problem
How is a dichotomous key made?

Objectives
In this BioLab, you will:
- **Classify** organisms on the basis of structural characteristics.
- **Develop** a dichotomous key.

Materials
sample keys from guidebooks
metric ruler

Skill Handbook
Use the **Skill Handbook** if you need additional help with this lab.

Procedure

1. Study the drawings of beetles on page 475 of your text.
2. Choose one characteristic of the beetles and classify the beetles into two groups based on that characteristic. Take measurements if you wish.
3. Record the chosen characteristic in a diagram like the one shown. Write the numbers of the beetles in each group on your diagram.
4. Continue to form subgroups within your two groups based on different characteristics. Record the characteristics and numbers of the beetles in your diagram until you have only one beetle in each group.
5. Using the diagram you have just made, make a dichotomous key for the beetles. Remember that each numbered step should contain two choices for classification. Begin with 1A and 1B. For help, examine sample keys provided by your teacher.
6. Exchange dichotomous keys with another team. Use their keys to identify the beetles.

Analyze and Conclude

1. **Comparing and Contrasting** Was the dichotomous key you constructed exactly like those of the other students? Explain your answer.
The keys may or may not have been alike. The groups may have first divided the beetles into groups based on different features, such as size rather than color.

Name Date Class

INVESTIGATE BioLab

Making a Dichotomous Key, *continued*

Chapter 17

2. **Analyzing Data** What characteristics were most useful for making a classification key for beetles? What characteristics were not useful?
Useful: size, color, and shape of various body parts, number of body sections, and antennae features; not useful: number of legs, number of antennae, habitat.

3. **Thinking Critically** Why do keys typically offer only two choices and not more?
Having only two choices makes it easy to analyze organisms. In many keys, the choice is that the organism either has or does not have a particular characteristic.

Data and Observations: Have students exchange their keys. If another group can use them, their accuracy is confirmed.

Name Date Class

MiniLab 18-1 Measuring a Virus

Measuring in SI

Can you use a light microscope to view a virus? Find out by measuring the size of a poliovirus in the photo on page 490 of your text and then comparing it to 0.2 μm, the size limit for viewing objects with a light microscope.

Procedure

1. Use the data table below.

Data Table

Values to measure and calculate	Measurement
Length of photo line in mm	
Diameter of poliovirus in mm	
Diameter of poliovirus in μm	

2. Examine the photo on page 490 of your text showing many polioviruses. The horizontal line you see would measure only 0.4 micrometer (μm) in length if the photo was not magnified 172 500×. Use this line for reference.
3. Calculate the diameter of one poliovirus. First, measure the length of the reference line in millimeters. Record the value in the table. Then, measure the diameter of a poliovirus in millimeters. Record the value in the table.
4. Use the following equation to calculate the actual diameter of the poliovirus (*X*). Record your answer in the table.

$$\frac{\text{photo line length in mm } (A)}{\text{diameter of virus in mm } (B)} = \frac{0.4\ \mu\text{m}}{\text{diameter of virus in } \mu\text{m } (X)}$$

Expected Results: Students should calculate the size of the poliovirus as follows.

$$\frac{2 \times 69\ \text{mm}}{5\ \text{mm}} = \frac{1\ \mu\text{m}}{(x)} = 0.036\ \mu\text{m}$$

Analysis

1. Explain why you cannot see viruses with a light microscope. Use specific numbers in your answer.
   The virus's diameter measures 0.036 μm, less than the 0.2 μm light microscope limit.
2. A bacterial cell may be 10 μm in size. How many polioviruses could fit across the top of such a bacterium?
   280 viruses (10 μm divided by 0.036 μm)

Name Date Class

MiniLab 18-2 Bacteria Have Different Shapes

Observing

Bacteria come in three different shapes: spherical (coccus), rodlike (bacillus), and spiral shaped (spirillum). They may appear singly or in pairs, chains, or clusters. Each species has a typical shape and reaction to Gram stain.

Procedure

1. Obtain slides of bacteria from your teacher.
2. Using low power, locate bacteria of one shape. Switch to high power. Look for individual cells and observe their shape. Observe also the size of the cells and their color. Then look for groups of bacterial cells to determine their arrangement. **CAUTION:** ***Use caution when working with a microscope and microscope slides.***
3. Repeat step 2 for bacteria with the other shapes. Then, compare the sizes of the bacteria.
4. Draw a diagram of each type of bacteria.

Expected Results: Students will observe spherical, rod, and spiral shaped bacteria.

Analysis

1. How do the sizes of the three bacteria compare?
   The spherical bacteria were smallest. The spirilla were most likely largest.
2. Which of the bacteria were Gram negative?
   the pink bacteria
3. What adaptive advantage might there be for bacteria to form groups of cells?
   exchange information and nuclear material

Name Date Class

DESIGN YOUR OWN BioLab

How sensitive are bacteria to antibiotics?

Chapter 18

PREPARATION

Problem
How can you determine which antibiotic most effectively kills specific bacteria?

Hypotheses
Decide on one hypothesis that you will test. Your hypothesis might be that the antibiotic with the widest zone of inhibition most effectively kills bacteria.

Objectives
In this BioLab, you will:

- **Compare** how effectively different antibiotics kill specific bacteria.
- **Determine** the most effective antibiotic to tread an infection that these bacteria might cause.

Possible Materials
cultures of bacteria
sterile nutrient agar petri dishes
antibiotic disks
sterile disks of blank filter paper
marking pen
long-handled cotton swabs
forceps
37°C incubator
metric ruler

Safety Precautions
Always wear goggles in the lab. Although the bacteria you will work with are not disease-causing, be careful not to spill them. Wash your hands with soap immediately after handling any bacterial culture. Carefully clean your work area after you finish. Follow your teacher's instructions about disposal of your swabs, cultures, and petri dishes.

Skill Handbook
Use the **Skill Handbook** if you need additional help with this lab.

Data and Observations: All antibiotic disks should show some inhibition. Results will vary depending on the antibiotic disks used. The untreated disks should show no zone of inhibition.

Name Date Class

DESIGN YOUR OWN BioLab

How sensitive are bacteria to antibiotics?, *continued*

Chapter 18

PLAN THE EXPERIMENT

1. Examine the materials provided by your teacher, and study the two photos. As a group, make a list of ways you might investigate your hypothesis.
2. Agree on one way that your group could investigate your hypothesis. Design an experiment in which you can collect quantitative data.
3. Make a list of numbered directions. In your list, include the amounts of each material you will need. If possible, use no more than one petri dish per person.
4. Design and construct a table for recording data. To do this, carefully consider what data you need to record and how you will measure the data. For example, how will you measure what happens around the antibiotic disks as the antibiotic diffuses into the agar?

Check the Plan
Discuss the following points with other group members to decide on your final procedure.

1. How you will set up your petri dishes. How many antibiotics can you test on one petri dish? How will you measure the effectiveness of each antibiotic? What will be your control?
2. Will you add the bacteria or the antibiotic disks first?
3. What will you do to prevent other bacteria from contaminating the petri dishes?
4. How often will you observe the petri dishes?
5. ***Make sure your teacher has approved your experimental plan before you proceed.***
6. Carry out your experiment. **CAUTION:** ***Wash your hands with soap and water after handling dishes of bacteria.***

ANALYZE AND CONCLUDE

1. **Measuring in SI** How did you measure the zones of inhibition? Why did you do it this way?
Most students will measure the diameter of the zone of inhibition and use the value as a basis of comparison.
2. **Drawing Conclusions** Suppose you were a physician treating a patient infected with these bacteria. Which antibiotic would you use? Why?
Use the one with the largest zone of inhibition because it inhibits bacterial growth the best.
3. **Analyzing the Procedure** What limitations does this technique have? If these bacteria were infecting a person, what other tests might increase your confidence about treating the person with the antibiotic that appears most effective against these bacteria?
The antibiotic may work in an agar culture, but not in human tissue. A doctor might also do blood tests to examine the kinds of white blood cells present, or look for the bacteria themselves.

Name Date Class

MiniLab 19-1 Observing Ciliate Motion

Observing and Inferring

The cilia on the surface of a paramecium move so that the cell normally swims through the water with one end directed forward. But when this end bumps into an obstacle, the paramecium responds by changing direction.

Procedure

1. Observe a *Paramecium* culture that has had boiled, crushed wheat seeds in it for several days.
2. Carefully place a drop of water containing wheat seed particles on a microscope slide. Gently add a coverslip.
3. Using low power, locate a paramecium near some wheat seed particles. **CAUTION:** ***Use caution when working with a microscope, glass slides, and coverslips.***
4. Watch the paramecium as it swims around among the particles. Record your observations of the organism's responses each time it contacts a particle.

Analysis

1. Describe what a paramecium does when it encounters an obstacle.
   It backs up, then proceeds forward in a new direction.
2. How long does the paramecium's response last?
   briefly
3. Describe any changes in the shape of the paramecium as it moved among the particles.
   The cell turned on its long axis, and it folded.

Expected Results: *Paramecium* typically reverses the direction in which its cilia are beating and backs away from a solid object that it contacts.

Name Date Class

MiniLab 19-2 Going on an Algae Hunt

Observing

Pond water may be teeming with organisms. Some are macroscopic organisms, but the majority are microscopic. Some may be heterotrophs, and others autotrophs. How can you tell them apart?

Procedure

1. Use the data table below.

Data Table

Diagram	Motile/Nonmotile	Unicellular/Multicellular
	Students will	**observe a variety of organisms.**

2. Place a drop of pond water onto a glass slide and add a coverslip. **CAUTION:** ***Use caution when working with a microscope, glass slides, and coverslips.***
3. Observe the pond water under low magnification of your microscope, and look for algae that may be present. Algae from a pond will usually be green or yellow-green in color.
4. Diagram several different species of algae in your data table and indicate if each is motile or nonmotile. Indicate if the algae are unicellular or multicellular.

Analysis

1. What characteristic distinguished algae from any protozoans that may have been present?
   Algae are green in color.
2. Explain how the characteristic in question 1 categorizes algae as autotrophs.
   Autotrophs contain chlorophyll for photosynthesis.
3. Did you observe any relationship between movement and size? Explain your answer.
   The smaller the algae the more likely they are motile.

Name Date Class

DESIGN YOUR OWN BioLab

How do *Paramecium* and *Euglena* respond to light?

Chapter 19

Preparation

Problem
Do both *Paramecium* and *Euglena* respond to light and do they respond in different ways? Among your group, decide on one type of protist activity that would constitute a response to light.

Hypotheses
Decide on one hypothesis that you will test. Your hypothesis might be that *Paramecium* will not respond to light and *Euglena* will respond, or that *Paramecium* will move away from light and *Euglena* will move toward light.

Objectives
In this BioLab, you will:
- **Prepare** slides of *Paramecium* and *Euglena* cultures and observe swimming patterns in the two organisms.
- **Compare** how these two different protists respond to light.

Possible Materials
Euglena culture
Paramecium culture
microscope
microscope slides
dropper
methyl cellulose
coverslips
metric ruler
index cards
scissors
toothpicks

Safety Precautions
Always wear goggles in the lab. Use caution when working with a microscope, glass slides, and coverslips. Wash your hands with soap and water immediately after working with protists and chemicals.

Skill Handbook
Use the **Skill Handbook** if you need additional help with this lab.

Plan the Experiment

1. Decide on an experimental procedure that you can use to test your hypothesis.
2. Record your procedure, step-by-step, and list the materials you will be using.
3. Design a data table in which to record observations and results.

Check the Plan
Discuss the following points with other group members to determine your final procedure.

1. What variables will you have to measure?
2. What will be your control?
3. What will be the shape of the light-controlled area(s) on your microscope slide?
4. Decide who will prepare materials, make observations, and record data.
5. ***Make sure your teacher has approved your experimental plan before you proceed further.***
6. To mount drops of *Paramecium* culture and *Euglena* culture on microscope slides, use a toothpick to place a small ring of methyl cellulose on a clean microscope slide. Place a drop of *Paramecium* or *Euglena* culture within this ring. Place a coverslip over the ring and culture. The thick consistency of methyl cellulose should slow down the organisms for easy observation.
7. Make preliminary observations of swimming *Paramecia* and *Euglena*. Then think again about the observation times that you have planned. Maybe you will decide to allow more or less time between your observations.
8. Carry out your experiment.

Name Date Class

DESIGN YOUR OWN BioLab

How do *Paramecium* and *Euglena* respond to light?, *continued*

Chapter 19

Analyze and Conclude

1. **Checking Your Hypothesis** Did your data support your hypothesis? Why or why not?
Answers may vary. Data must be used to either support or reject student hypotheses.

2. **Comparing and Contrasting** Compare and contrast the responses of *Paramecium* and *Euglena* to light and darkness. What explanations can you suggest for their behavior?
***Euglena* are attracted to light. Paramecium avoid bright light. *Euglena* need light to make food. Heterotropohic, *Paramecium* can find food in dim and dark places.**

3. **Making Inferences** Can you use your results to suggest what sort of responses to light and darkness you might observe using other heterotrophic or autotrophic protists?
Most autotrophs show a positive response to light whereas most heterotrophs show a negative response.

Data and Observations: Data should indicate that *Euglena* are attracted to light; *Paramecium* are not.

Name Date Class

MiniLab 20-1 Growing Mold Spores

Observing and Inferring

Any mold spore that lands in a favorable place can germinate and produce hyphae. Can you identify a condition necessary for the growth of bread mold spores?

Procedure

1. Place two slices of freshly baked bakery bread on a plate. Sprinkle some water on one slice to moisten its surface. Leave both slices uncovered for several hours.
2. Sprinkle a little more water on the moistened slice, and place both slices in their own plastic, self-seal bags. Trap air in each bag so that the plastic does not touch the bread's surface. Then seal the bags and place them in a darkened area at room temperature.
3. After five days, remove the bags and look for mold.
4. Remove a small piece of mold with a forceps, place it on a slide in a drop of water, and add a coverslip. Observe the mold under a microscope's low power and high power. **CAUTION:** ***Use caution when working with a microscope, glass slides, and coverslips. Wash your hands with soap and water after working with mold. Dispose of the mold as your teacher directs.***

Analysis

1. Did you observe mold growth on the moistened bread? On the dry bread? How does this experiment demonstrate that there are mold spores in your classroom?
   Only moist bread had mold. It suggests that there might be spores in the classroom air.
2. What conclusions can you draw about the conditions necessary for the growth of a bread mold?
   Molds require water and food for growth.

Expected Results: After 4–6 days, mold will grow only on moist bread slices.

Name Date Class

MiniLab 20-2 Examining Mushroom Gills

Classifying

Spore prints can often help in mushroom identification by revealing the pattern of a mushroom's gills and the color of its spores. Use this technique to see how a mushroom's gills are arranged.

Procedure

1. Break off the stalks from several grocery-store mushrooms. Place the caps in a paper bag for a few days.
2. When the undersides of the caps are very dark brown, set the caps, gill side down, on a white sheet of paper. Be sure that the gills are touching the surface of the paper.
3. After leaving the caps undisturbed overnight, carefully lift the caps from the paper and observe the results.
4. Wash your hands with soap and water. Dispose of fungi as your teacher directs.

Analysis

1. What color are the spores on the paper?
   brown
2. How does the pattern of spores on the paper compare with the arrangement of gills on the underside of the mushroom cap that produced it?
   Spores are formed toward the edges of the gills. The patterns are the same.

Expected Results: Spores will form a pattern on the paper corresponding to the gills' locations.

Name Date Class

INTERNET BioLab

Does temperature affect the metabolic activity of yeast?

Chapter 20

Preparation

Problem

How can you determine the affect of temperature on the metabolism of yeast? Brainstorm ideas among the members of your group.

Hypotheses

Decide on one hypothesis that you will test. Your hypothesis might be that low temperature slows down the metabolic activity of yeast, or that high temperature speeds up the metabolic activity of yeast.

Objectives

In this BioLab, you will:

- **Measure** the rate of yeast metabolism using a BTB color change as a rate indicator.
- **Compare** the rates of yeast metabolism at several temperatures.
- **Use the Internet** to collect and compare data from other students.

Possible Materials

bromothymol blue solution (BTB)
straw
small test tubes (4)
large test tubes (3)
one-hole stoppers with glass tube inserts for large test tubes (3)
yeast/white corn syrup mixture
water/white corn syrup mixture
water/yeast mixture
test-tube rack
250 mL beakers (3)
ice cubes
Celsius thermometer
hot plate
50 mL graduated cylinder
glass-marking pencil
10 cm rubber tubing (3)
aluminum foil

Safety Precautions

Always wear goggles in the lab. Be careful in attaching rubber tubing to the glass tube inserts in the stoppers. Avoid touching the top of the hot plate. Wash your hands thoroughly after cleaning out test tubes at the end of your experiments.

Skill Handbook

Use the **Skill Handbook** if you need additional help with this lab.

Data and Observations: More carbon dioxide gas is produced by yeast immersed in warm water temperatures. These tubes change color more rapidly than those at cooler temperatures.

Name Date Class

INTERNET BioLab

Does temperature affect the metabolic activity of yeast?, *continued*

Chapter 20

Plan the Experiment

1. Decide on ways to test your group's hypothesis.
2. Record your procedure, and list the materials and amounts of solutions that you will use.
3. Design a data table for recording your observations.
4. Pour 5 mL of BTB solution into a test tube. Use a straw to blow gently into the tube until you see a series of color changes. Cover this tube with aluminum foil, and set it aside in a test-tube rack. Record your observations of the color changes caused by carbon dioxide in your breath.

Check the Plan

Discuss the following to decide on your procedure.

1. What data on color change and time will you collect? How will you record your data?
2. What variables will you control?
3. What control will you use?
4. Assign tasks for each member of your group.
5. ***Make sure your teacher has approved your experimental plan before you proceed further.***
6. Carry out your experiment.
7. Visit the Glencoe Science Web Site at **www.glencoe.com/sec/science** to **post your data.**

Analyze and Conclude

1. **Checking Your Hypotheses** Explain whether your data support your hypothesis. Use your experimental data to support or reject your hypothesis concerning temperature effects on the rate of yeast metabolism.
   Answers will vary. The warmer the water, the more rapid the color change of BTB.
2. **Using the Internet** Did the data from the Internet support your hypothesis? Explain your answer.
   The Internet data should support student hypotheses.
3. **Making Inferences** What must be the role of white corn syrup in this experiment?
   It is the food for yeast.
4. **Identifying Variables** Describe some variables that had to be controlled in this experiment. Explain how you controlled each variable.
   The volume of water/corn syrup and the amount of yeast used were variables and had to be kept constant.
5. **Drawing Conclusions** Describe the control used in your experiment and how your experimental results enabled you to draw conclusions about the effect of temperature on yeast metabolism. Did your experiment clearly show that differences in rates of yeast metabolism were due to temperature differences?
   Controls will vary, such as a tube with no yeast/corn syrup. Some data will show more difference than other data.

Name Date Class

MiniLab 21-1

Applying Concepts

Examining Land Plants

Liverworts are considered to be one of the simplest of all land plants. They show many of the adaptations that other land plants have evolved that enable them to survive on a land environment.

Procedure

1. Examine a living or preserved sample of Marchantia. **CAUTION:** ***Wear disposable latex gloves when handling preserved materials.***
2. Note and record the following observations. Is the plant unicellular or multicellular? Does it have a top and bottom? How do these differ? Is it one cell in thickness or many cells thick? Does the plant seem to grow upright like a tree or is it flat to the ground?
3. Use a dissecting microscope to examine its top and bottom surfaces. Are tiny holes or pores present? If you answer "yes," which surface has pores?

Analysis

1. How might having a multicellular, thick body be an advantage to life on land?
   Water can be retained in cells that help the plant retain moisture.
2. Are any structures present that resemble roots? How might they help a land plant?
   yes; enables plant to obtain water and minerals from soil, helps anchor plant to ground
3. How might its growth pattern help it survive land?
   Growing close to ground may reduce drying out from wind and help the plant retain moisture.
4. What might be the role of any pores observed on the plant? Why is the location of the pores critical to surviving on a land environment?
   gas exchange—carbon dioxide and oxygen can enter plant; carbon dioxide and oxygen are present in air—upper surface is exposed to air

Expected Results: Students will examine a liverwort and correlate their observations on liverwort structure with adaptations of land plants.

Name Date Class

MiniLab 21-2

Comparing and Contrasting

Looking at Modern and Fossil Plants

Many modern-day plants have relatives that are known only from the fossil record. Are modern-day plants similar to their fossil relatives? Are there any differences?

Procedure

1. Examine a preserved or living sample of *Lycopodium*, a club moss. **CAUTION:** ***Wear disposable latex gloves when handling preserved material.***
2. Note and record the following observations:
   a. Does the plant grow flat or upright like a tree?
   b. Describe the appearance of its leaves and its stem.
   c. Measure the plant's height and diameter in centimeters.
3. Diagram A on page 586 of your text is a representation of a fossil relative called *Lepidodendron*. Record the same observations (a–c).
4. Repeat steps a–c only this time use a preserved or living sample of *Equisetum*, a horsetail. Compare it to Diagram B on page 586 of your text, a representation of a fossil relative called *Calamites*.

Expected Results: Students will find a number of similarities and some differences between the living plants and their fossil relatives. Numbers of similarities and differences will depend on the thoroughness of student observations.

Analysis

1. Describe the similarities and differences between *Lycopodium* and *Lepidodendron*. Do your observations justify their closeness as relatives? Explain.
   Similarities—green, grow upright, have scalelike leaves. Differences—tall and thick fossil stem, short and thin living stem, leaves on top of fossil stem, ridged surface of fossil stem, leaves on living stem. Student answers on the closeness of the relationship may vary. If going by general appearance, students may agree that they are closely related. If using the total number of similarities and differences, students may disagree.
2. Describe the similarities and differences between *Equisetum* and *Calamites*. Do your observations justify their closeness as relatives? Explain.
   Similarities—green, grow upright, leaves are in whorls, leaves are small, rings on stem. Differences—fossil has branches while living is one single stem, fossil is taller and thicker, living plant leaves appear to be much smaller than fossil. Student answers on the closeness of the relationship may vary. If going by general appearance, students may agree that they are closely related. If using the total number of similarities and differences, students may disagree.

Name Date Class

INTERNET BioLab: Researching Trees on the Internet

Chapter 21

Preparation

Problem
Use the Internet to find different trees that would be suitable for planting in your community.

Objectives
In this BioLab, you will:
- **Research** the characteristics of five different trees.
- **Use the Internet** to collect and compare data from other students.
- **Conclude** which trees would be most suitable for planting in your community.

Materials
Internet access

Skill Handbook
Use the **Skill Handbook** if you need additional help with this lab.

Procedure

1. Use the data table.
2. Pick five trees that you wish to research. Note: Your teacher may provide you with suggestions if necessary.
3. Go to the Glencoe Science Web Site at **www.glencoe.com/sec/science** to find links to information needed for this BioLab.
4. Record the information in your data table.

Data Table

	1	2	3	4	5
Tree Name (common name)					
Scientific Name					
Kingdom					
Division					
Soil/Water Preference					
Temperature Tolerance					
Height at Maturity					
Speed of Growth					
General Shape					
Disease Resistance/Problems					
Special Care					
Leaf Shape					
Shade Provider					
Additional Information					

Name Date Class

INTERNET BioLab: Researching Trees on the Internet, *continued*

Chapter 21

Analyze and Conclude

1. **Defining Operationally** Explain the difference between trees classified as either Coniferophyta or Anthophyta.
   Coniferophytes are cone-bearing trees, seeds formed on cones, evergreens; anthophytes are flowering plants, seed formed within a fruit, usually deciduous.
2. **Analyzing** Was the information provided on the Internet helpful in completing your data table? Explain your answer.
   Student answers will vary. The Internet (if used) will provide information for the data table.
3. **Thinking Critically** What do you consider to be the most important characteristic when deciding on the most suitable tree for your community? Explain your answer.
   Student answers will vary. Students may select speed of growth, temperature tolerance, or disease resistance as the most important qualities.
4. **Using the Internet** Using the information you gathered from the Internet, which tree species would most likely be the:
   a. most suitable for your community? Explain your answer.
   Student answers will vary.
   b. least suitable for your community? Explain your answer.
   Student answers will vary.
5. **Applying** Explain why tree selections would differ if your community were located in:
   a. Tucson, Arizona
   High summer temperature and drought tolerance would be important factors.
   b. Los Angeles, California
   Long periods of drought during the summer would be an important factor.
   c. Fairbanks, Alaska
   Temperature tolerance during the winter would be an important factor.

Data and Observations: Student data tables will vary depending on the initial trees

Name Date Class

MiniLab 22-1

Experimenting

Identifying Fern Sporangia

When you admire a fern growing in a garden or forest, you are admiring the plant's sporophyte generation. Upon further examination, you should be able to see evidence of spores being formed. Typically, the evidence you are looking for can be found on the underside of the ferns' fronds.

Procedure

1. Place a drop of water and a drop of glycerin at opposite ends of a glass slide.
2. Use forceps to gently pick off one sorus from a frond. Place it in the drop of water and add a coverslip.
3. Add a second sorus to the glycerin and add a coverslip.
4. Observe both preparations under low-power magnification and note any similarities and differences. Look for large sporangia (resembling heads on a stalk) and spores (tiny round bodies released from a sporangium). **CAUTION:** ***Use caution when working with a microscope, microscope slides, and coverslips.***

Analysis

1. Were spores more visible in water or in glycerin?
   glycerin
2. What did the glycerin do to the sporangium?
   Glycerin caused the sporangia to break open.
3. Formulate a hypothesis that may explain how sporangia naturally burst.
   Student answers may vary—drying out of sporangium, absorption of water.
4. Formulate a hypothesis that may explain how sporangia were affected by glycerin.
   Student answers may vary—glycerin causes osmotic imbalance.

Expected Results: Students will see a few spores released from sporangia under water but should see many more in glycerin.

Name Date Class

MiniLab 22-2

Comparing and Contrasting

Comparing Seed Types

Anthophytes are classified into two classes, the monocotyledons (monocots) and dicotyledons (dicots) based on the number of seed leaves.

Procedure

1. Use the data table below.
2. Examine the variety of seeds given to you. Use forceps to gently remove the seed coat or covering from each seed if one is present.
3. Determine the number of cotyledons present. If two cotyledons are present, the seed will easily separate into two equal halves. If one cotyledon is present, it will not separate into halves. Record your observations in the data table.
4. Add a drop of iodine stain to rice and a lima bean seed. Note the color change. **CAUTION:** ***Wash your hands with soap and water after handling chemicals.*** Record your observations in the data table.

Data Table

Seed name	Number of cotyledons	Monocot or dicot	Color with iodine
Lima bean			
Rice			
Pea			—
Rye			—

Analysis

1. Starch turns purple when iodine is added to it. Describe the color change when iodine was added to rice and lima bean seeds.
   The seeds turn purple.
2. Hypothesize why seeds contain stored starch.
   Starch supplies a growing embryo with food.

Expected Results: The lima bean, pea, and sunflower are dicots. All others are monocots. Rice and lima bean seeds will turn blue.

Name Date Class

Design Your Own BioLab

How can you make a key for identifying conifers?

Chapter 22

Preparation

Problem

What kinds of characteristics can be used to create a key for identifying different kinds of conifers?

Hypotheses

State your hypothesis according to the kinds of characteristics you think will best serve to distinguish among several conifer groups. Explain your reasoning.

Objectives

In this BioLab, you will:

- **Compare** structures of several different conifer specimens.
- **Identify** which characteristics can be used to distinguish one conifer from another.
- **Communicate** to others the distinguishing features of different conifers.

Possible Materials

twigs, branches, and cones from several different conifers that have been identified for you

Safety Precautions

Always wash your hands after handling biological materials. Always wear goggles in the lab.

Skill Handbook

Use the **Skill Handbook** if you need additional help with this lab.

Plan the Experiment

1. Make a list of characteristics that could be included in your key. You might consider using shape, color, size, habitat, or other factors.
2. Determine which of those characteristics would be most helpful in classifying your conifers.
3. Determine in what order the characteristics should appear in your key.
4. Decide how to describe each characteristic.

Check the Plan

1. The traits described at each fork in a key are often pairs of contrasting characteristics. For example, the first fork in a key to conifers might compare "needles grouped in bundles" with "needles attached singly."
2. Someone who is not familiar with conifer identification should be able to use your key to correctly identify any conifer it includes.
3. ***Make sure your teacher has approved your experimental plan before you proceed further.***
4. Carry out your plan by creating your key.

Name Date Class

Design Your Own BioLab

How can you make a key for identifying conifers?, *continued*

Chapter 22

Analyze and Conclude

1. **Checking Your Hypothesis** Have someone outside your lab group try using your key to identify your conifer specimens. If they are unable to make it work, try to determine where the problem is and make improvements. **(By using student key design on the overhead, the entire class can determine whether or not the key works.)**
2. **Making Inferences** Is there only one correct way to design a key for your specimens? Explain why or why not.
   Student key designs placed on the overhead will illustrate the diversity of student trait choices used to organize and design each key.
3. **Relating Concepts** Give one or more examples of situations in which a key would be a useful tool.
   It can distinguish poisonous from nonpoisonous plants, and harmful from nonharmful insects.

Data and Observations: Have students record their keys and turn them in at the end of class. Make transparencies of sample keys and use them in class the following day as a means for illustrating correct and incorrect key design.

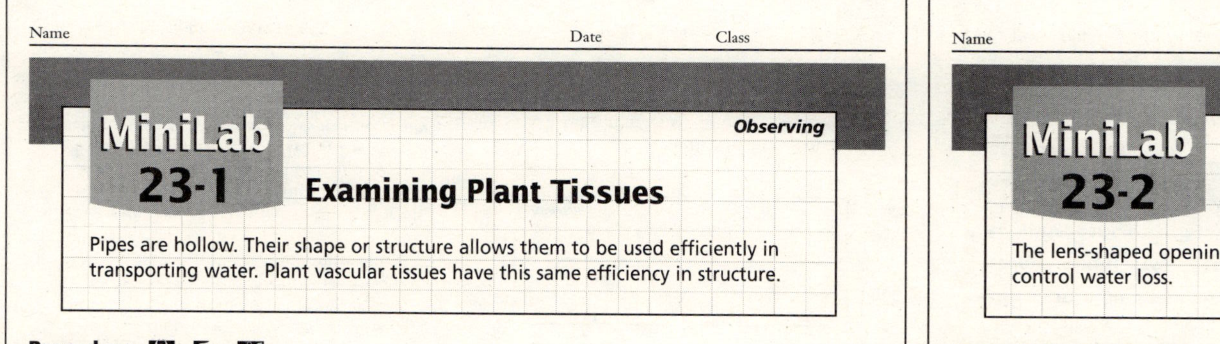

Name Date Class

MiniLab 23-1 Examining Plant Tissues

Observing

Pipes are hollow. Their shape or structure allows them to be used efficiently in transporting water. Plant vascular tissues have this same efficiency in structure.

Procedure

1. Snap a celery stalk in half and remove a small section of "stringy tissue" from its inside.
2. Place the material on a glass slide. Add several drops of water. Place a second glass slide on top. **CAUTION:** ***Use caution when working with a microscope and slides.***
3. Press down evenly on the top glass slide with your thumb directly over the plant material.
4. Remove the top glass slide. Add more water if needed. Add a coverslip.
5. Examine the celery material under low- and high-power magnification. Diagram what you see.
6. Repeat steps 2–5 using some of the soft tissue inside the celery stalk.

Analysis

1. Describe the appearance of the stringy tissue inside the celery stalk. What may be the function of this tissue?
   tubelike, pipelike, stringlike, nongreen or colorless; transport of materials
2. Describe the appearance of the soft tissue inside the celery stalk. What may be the function of this tissue?
   cubelike, square cells, green in color; storage
3. Does the structure of these tissues suggest their functions?
   Yes. Long, narrow cells are ideal for transporting water and nutrients. Cubelike cells would be most suitable for food storage.

Expected Results: The stringy tissue will appear as nongreen tissue that resembles a track, or roadway. The soft tissue will appear green and cubelike in shape.

Name Date Class

MiniLab 23-2 Looking at Stomata

Observing

The lens-shaped openings in the epidermis of a leaf allow gas exchange and help control water loss.

Procedure

1. Make a wet mount by tearing a leaf at an angle to expose a thin section of epidermis. Use tap water to make wet mounts of both the upper and lower epidermis.
2. Examine each of your slide preparations under the microscope. Draw or take down a written description of what you see. **CAUTION:** ***Use caution when working with a microscope, microscope slides, and coverslips.***
3. Make another wet mount using a 5 percent salt solution instead of tap water. Examine the slide under the microscope and record your observations.

Analysis

1. What do the cells of the leaf epidermis look like? What is their function?
   Cells look like interlocked puzzle pieces. The cells are protective.
2. How do the epidermal cells differ from guard cells? Which cells, if any, contain chloroplasts?
   Guard cells are sausage-shaped and have chloroplasts. Epidermal cells do not have chloroplasts and are irregular in shape.
3. What differences in the stomata did you notice when you used a salt solution to prepare your wet mount? Can you explain what happened in terms of osmosis?
   Stomata are closed in the salt solution. The higher water concentration inside the cells compared with that outside of the cells caused water to move out of the guard cells. This causes the guard cells to collapse and close the stomata.

Expected Results: Chloroplasts will be seen only in guard cells. The saltwater mount will show closed guard cells. Thus, stomata will appear closed in comparison to the plain water wet mount.

INVESTIGATE BioLab

Determining the Number of Stomata on a Leaf

Chapter 23

Preparation

Problem
How can you count the total number of stomata on a leaf?

Objectives
In this BioLab, you will:
- **Measure** the area of a leaf.
- **Observe** the number of stomata seen under a high-power field of view.
- **Calculate** the total number of stomata on a leaf.

Materials
microscope
glass slide
water and dropper
green leaf from an onion plant
single-edged razor blade
ruler
glass cover

Safety Precautions
Wear latex gloves when handling an onion.

Skill Handbook
Use the **Skill Handbook** if you need additional help with this lab.

Procedure

1. Use Data Tables 1 and 2.
2. Obtain an onion leaf and carefully cut it open lengthwise using a single-edged razor blade. CAUTION: *Be careful when cutting with a razor blade.*
3. Measure the length and width of your onion leaf in millimeters. Record these values in Data Table 2.
4. Remove a small section of leaf and place it on a glass slide with the dark green side facing DOWN.
5. Add several drops of water and gently scrape away all green leaf tissue using a back and forth motion with the razor blade. An almost transparent layer of leaf epidermis will be left on the slide.
6. Add water and a cover glass to the epidermis and observe under low-power magnification. Locate an area where guard cells and stomata can clearly be seen. CAUTION: *Use caution when working with a microscope, microscope slides, and coverslips.*
7. Switch to high-power magnification.
8. Count and record the number of stomata in your field of view. Consider this trial 1. Record your count in Data Table 1.

Data Table 1

Trial	Number of stomata
1	
2	
3	
4	
5	
Total	
Average	

9. Move the slide to a different area. Count and record the number of stomata in this field of view. Consider this trial 2.

INVESTIGATE BioLab

Determining the Number of Stomata on a Leaf, *continued*

Chapter 23

10. Repeat step 9 three more times. Calculate the average number of stomata observed in a high-power field of view.
11. Calculate the total number of stomata on the entire onion leaf by following the directions in Data Table 2.

Data Table 2

Length of leaf portion in mm	= ________ mm
Width of leaf portion in mm	= ________ mm
Calculate area of leaf (length × width)	= ________ mm²
Calculate number of high-power fields of view on leaf (area of leaf ÷ 0.07 mm², the area of one high-power field of view)	= ________
Calculate total number of stomata (number of high-power fields of view × average number of stomata per high-power field of view from Data Table 1)	= ________

Analyze and Conclude

1. **Communicating** Compare your data with those of your other classmates. Offer several reasons why your total number of stomata for the leaf may not be identical to your classmates'.
   mathematical errors, different average numbers of stomata, different leaf sizes
2. **Thinking Critically** Analyze the following steps of this experiment and explain how you can change the procedure to improve the accuracy of your data.
   a. five trials in Data Table 1
   increase the number of samples
   b. using 0.07 mm^2 as the area of your high-power field of view
   calculate microscope's high-power area
3. **Concluding** Would you expect all plants to have the same number of stomata per high-power field of view? Explain your answer.
   Different plants will have different numbers of stomata per high-power field; stomate numbers are characteristic for specific plant species.
4. **Comparing and Contrasting** What are the advantages to using sampling techniques? What are some limitations?
   Saves time and energy; you do not get an actual or true count, only an approximation.

Data and Observations: Student answers will vary. The number of stomata will be several thousand.

Name Date Class

MiniLab 24-1 Growing Plants Asexually

Experimenting

Plants are capable of reproducing asexually. Reproductive cells such as egg or sperm are not needed in asexual reproduction. Plants are able to use structures such as roots, stems, and even leaves to produce new offspring.

Procedure

1. Prepare three different plant parts for study using diagrams A, B, and C on page 654 of your text as a guide.
2. Observe any changes that occur to your plants over the next two weeks.
3. Design a data table that will provide enough room for diagrams of your observations. The number of days since the start of the experiment should be included.
4. Make your initial diagrams of the plant parts today and label these diagrams as "Day 1."
5. Observations should be made every three days. Replace any lost water as needed.

Analysis

1. What experimental evidence do you have that:
   a. plants use a variety of structures for asexual reproduction?
   The experimental procedure demonstrates that different plant parts can generate new growth.
   b. asexual reproduction is a rapid process?
   The appearance of new growth occurred within several days.
   c. asexual reproduction requires only one parent?
   Only one plant was used for each experimental setup.
2. Describe several advantages of asexual reproduction in plants.
   Student answers may vary; they may include faster growth and that all offspring are identical to parent.

Expected Results: All plant tissues will produce new growth. The garlic clove will show new root and stem/leaf growth. The carrot will show new leaves. The potato will show new stems and leaves.

Name Date Class

MiniLab 24-2 Looking at Germinating Seeds

Observing

Seeds are made up of a plant embryo, a seed coat, and in some plants, a food-storage tissue. Monocot and dicot seeds differ in their internal structures.

Procedure

1. Obtain from your teacher a soaked, ungerminated corn kernel (monocot), a bean seed (dicot), and corn and bean seeds that have begun to germinate.
2. Remove the seed coats from each of the ungerminated seeds, and examine the structures inside. Use low-power magnification. Locate and identify each structure of the embryo and any other structures you observe.
3. Examine the germinating seeds. Locate and identify the structures you observed in the dormant seeds.

Analysis

1. Diagram the dormant embryos in the soaked seeds, and label their structures.
   Labels should include cotyledon(s), embryo.
2. Diagram the germinating seeds, and label their structures.
   Diagrams of dicot should include cotyledons, embryo, epicotyl, radicle, plumule, and hypocotyl. Diagrams of monocot should include only cotyledon and embryo.
3. List at least three major differences you observed in the internal structures of the corn and bean seeds.
   Monocot has one cotyledon, small embryo, is slow to germinate, and its embryo parts are hard to differentiate. Dicot seed has two cotyledons, larger embryo, and easily seen embryo parts.

Expected Results: Germinating seeds have larger embryos.

Name Date Class

INVESTIGATE BioLab

Examining the Structure of a Flower

Chapter 24

Preparation

Problem

What do the parts of a flower look like? How are they arranged?

Objectives

In this BioLab, you will:

- **Observe** the structures of a flower.
- **Identify** the functions of flower parts.

Materials

flower—any complete flower that is available locally, such as phlox, lily, or tobacco flower
hand lens (or stereomicroscope)
colored pencils (red, green, blue)
2 microscope slides
water
dropper
2 coverslips
microscope
single-edged razor blade

Safety Precautions

Always wear goggles in the lab. Handle the razor blade with extreme caution. Always cut away from you. Use caution when working with a microscope and slides. Wash your hands with soap and water after handling plant material.

Skill Handbook

Use the **Skill Handbook** if you need additional help with this lab.

Procedure

1. Examine your flower. Locate the sepals and petals. Note their numbers, size, color, and arrangement on the flower stem.
2. Remove the sepals and petals from your flower by gently pulling them off the stem. Locate the stamens, each of which consists of a thin filament with a pollen-filled anther on the tip. Note the number of stamens.
3. Locate the pistil. The stigma at the top of the pistil is often sticky. The style is a long, narrow structure that leads from the stigma to the ovary.
4. Place an anther from one of the stamens onto a microscope slide and add a drop of water. Cut the anther into several pieces with the razor blade. **CAUTION:** ***Always take care when using a razor blade.***
5. Examine the anther under low and high power of your microscope. The small, dotlike structures are pollen grains.
6. Slice the ovary in half lengthwise with the razor blade. Mount one half, cut side facing up, on a microscope slide.
7. Examine the ovary section with a hand lens or stereomicroscope. The many small, dotlike structures that fill the two ovary halves are ovules. Each ovule contains an egg cell that is not visible under low power. A tiny stalk connects each ovule to the ovary wall.
8. Identify the ovary and ovules.
9. Make a diagram of the flower, labeling all its parts. Color the female reproductive parts red. Color the male reproductive parts green. Color the remaining parts blue.

Name Date Class

INVESTIGATE BioLab

Examining the Structure of a Flower, *continued*

Chapter 24

Analyze and Conclude

1. **Observing** How many stamens are present in your flower? How many pistils, ovaries, sepals, and petals?
   Stamens, petals, and sepals will be in the multiples of 3 if a monocot, multiples of 4 or 5 if a dicot. One pistil and ovary will be present.
2. **Comparing and Contrasting** Make a reasonable estimate of the number of pollen grains in the anther and the number of ovules in an ovary of your flower.
   thousands of pollen grains, 10 to 100 ovules
3. **Interpreting Data** Which produces more? Pollen grains by one anther? Ovules produced by one ovary? Give a possible explanation for your answer.
   Yes; the flower increases the probability of a pollen cell landing on the correct stigma by producing and releasing a large number of pollen grains.

Data and Observations: Student diagrams, when colored properly, should show the petals and sepals as blue, pistil as red, and stamens as green.

Name Date Class

MiniLab 25-1 Observing Animal Characteristics

Observing and Inferring

Animals come in a variety of sizes and shapes, and can be found living in a number of different habitats.

Procedure

1. Use the data table below.
2. Add a few bristles from an old toothbrush to a glass slide. Add a drop of water containing rotifers to your slide. The drop should cover the bristles. Add a coverslip. **CAUTION:** ***Use caution when working with a microscope, slides, and coverslips.***
3. Observe your rotifers under low-power magnification.
4. Use the data table to record the characteristics that you were able to see. Describe the evidence for each trait.

Data Table

Animal Characteristic	Observed? (Yes or No)	Evidence
Multicellular		
Feeding		
Movement		
Size in mm		

Analysis

1. Are these organisms autotrophs or heterotrophs?
   They are not green, thus are probably not autotrophs.
2. Were you able to see evidence of feeding? Explain.
   Student answers will vary. Most rotifers will be actively feeding on bacteria; the wheel-like action of cilia implies that these animals are pulling water and food into their bodies.
3. Are rotifers multicellular? Explain.
   Students won't see individual cells, so they infer that the rotifers are multicellular.

Expected Results: Students will observe movement and feeding. They will find a size range from about 0.04 to 2.0 mm in length. Students will infer that these animals are multicellular.

Name Date Class

MiniLab 25-2 Check Out a Vinegar Eel

Observing and Inferring

Vinegar eels are roundworms with pseudocoeloms. They exhibit an interesting pattern of locomotion because they have only longitudinal (lengthwise) muscles.

Procedure

1. Prepare a wet mount of vinegar eels. **CAUTION:** ***Use caution when working with a microscope and slides.***
2. Observe them under low-power magnification.
3. Note their pattern of locomotion. Prepare a series of diagrams that illustrate their pattern of movement.
4. Time how fast they move by timing in seconds how long it takes for one roundworm to move across the center of your field of view. Find out the diameter of your low-power field in mm. Calculate vinegar eel speed in mm/sec. You may want to time several animals and average their speed.

Analysis

1. Name the type of symmetry present in vinegar eels.
   bilateral symmetry
2. Describe the pattern of locomotion for vinegar eels.
   series of jerky contractions
3. How does the pseudocoelom aid vinegar eels in locomotion?
   The body cavity if filled with fluid. Body muscles can brace and contract against this fluid.
4. What is the speed of locomotion for a vinegar eel? Based on the speed of your vinegar eel, predict the speed in mm/sec for a flatworm. Explain your answer.
   The speed will range between 2 and 5 mm/sec. The speed of a flatworm will be slower due to the fact that flatworms have no body cavity. A pseudocoelom allows more efficient muscle contraction.

Expected Results: Students will observe the movement of vinegar eels as a series of jerking contractions. The speed of these animals will be in the range of 2 to 5 mm/second.

Name Date Class

INTERNET BioLab

Zebra Fish Development

Chapter 25

PREPARATION

Problem
What do the developmental stages of the zebra fish look like?

Objectives
In this BioLab, you will:
- **Observe** stages of zebra fish development.
- **Record** all observations in a data table.
- **Use the Internet** to collect and compare data from other students.

Skill Handbook
Use the **Skill Handbook** if you need additional help with this lab.

Materials
aquarium
zebra fish
turkey baster
beaker
dropper
petri dish
binocular microscope
wax pencil or labels

Safety Precautions
Always wear safety goggles in the lab. Use caution when working with a binocular microscope and glassware.

PROCEDURE

1. Use the data table
2. Using the turkey baster, draw up water containing zebra fish embryos from the bottom of the aquarium.
3. Release the water into a beaker and allow the embryos to settle to the bottom.
4. Label a petri dish with your name and class period. Fill the bottom of the dish with aquarium water. Using a dropper, place several embryos in your dish.
5. Your teacher will advise you as to the approximate time that fertilization took place. All ages should be reported in your data table as hpf (hours past fertilization).
6. Using a binocular microscope, observe the embryos. Make a diagram in your data table of the embryos' appearances and indicate the age of the embryos in hpf.
7. Go to the Glencoe Science Web Site at **www.glencoe.com/sec/science** to **post your data.**
8. Continue to observe your embryos daily for a minimum of one week. Note the appearance of new organs and when movement is first seen. If you wish to continue watching developmental changes, consult with your teacher for directions. **CAUTION:** ***Wash your hands with soap and water immediately after completing observations.***

Data and Observations:
Between 1–2 hpf: dividing cells will appear above yolk sac.
Between 2–5 hpf: blastula formation.
Between 5–10 hpf: gastrulation formation is complete with appearance of somites.
Between 10–28 hpf: tissues and organs form.
Between 28–72 hpf: fins, gills, mouth form.

Name Date Class

INTERNET BioLab

Zebra Fish Development, *continued*

Chapter 25

Data Table

Date	hpf	Diagram	Observations

ANALYZE AND CONCLUDE

1. **Communicating** Explain why zebra fish are ideal animals for studying embryonic development.
   They are inexpensive, easy to care for, form many embryos, and show rapid development.
2. **Thinking Critically** Explain why you may not have been able to see stages such as a blastula or gastrula.
   The early stages occur rapidly after fertilization.
3. **Thinking Critically** Suggest how you could change the experiment's design to allow for observing these stages.
   Students could reset the timer, or use embryos that were collected and preserved by an earlier class.
4. **Using the Internet** Visit the Glencoe Science Web Site for links to Internet sites that will help you complete sequences of the major changes during development of zebra fish.
   a. between 1 and 10 hpf. Include labeled diagrams of these changes.
   b. between 10 and 28 hpf. Include labeled diagrams.
   c. between 28 and 72 hpf. Include labeled diagrams.

   For questions a–c, student answers should agree with the information in Data and Observations. Encourage students to print out Internet diagrams and include them in their answers to this question.

Name Date Class

MiniLab 26-1 Watching Hydra Feed

Observing

Hydras are freshwater cnidarians. They show the typical polyp body plan and symmetry associated with all members of this phylum. Observe how they capture their food.

Procedure

1. Use a dropper to place a hydra into a watch glass filled with water. Wait several minutes for the animal to adapt to its new surroundings. **CAUTION:** ***Use caution when working with a microscope and glassware.***
2. Observe the hydra under low-power magnification.
3. Formulate a hypothesis as to how this animal obtains its food and/or catches its prey.
4. Place brine shrimp in a culture dish of freshwater to avoid introducing salt into the watch glass.
5. Add a drop of brine shrimp to the watch glass while continuing to observe the hydra through the microscope.
6. Note which structures the hydra uses to capture food.

Analysis

1. Describe how the hydra captures food.
Nematocysts and tentacles work together to capture food.

2. Was your hypothesis supported or rejected?
Student answers will vary depending on their original hypothesis.

3. Sequence the events that take place when a hydra captures and feeds upon its prey.
prey organisms brush against tentacles, nematocysts are released, prey is captured by tentacles, tentacles push food into hydra's mouth

4. Explain how your observations support the fact that hydras have both nervous and muscular systems.
muscles move the tentacles, nervous system directs tentacles to trap food and push food into mouth

Expected Results: Students will see the tentacles of hydra surround and capture brine shrimp. Food is then pushed by the tentacles into the animals' mouth.

Name Date Class

MiniLab 26-2 Observing the Larval Stage of a Pork Worm

Observing

You can observe the larval stage of a pork worm (*Trichinella spiralis*) embedded within the muscle tissue of its host. It will look like a curled up hot dog surrounded by muscle tissue.

Procedure

1. Examine a prepared slide of pork worm larvae under the low-power magnification of your microscope.
2. Locate several larvae by looking for "spiral worms enclosed in a sac." All other tissue is muscle.
3. Estimate the size of the larva in µm.
4. Diagram one larva. Indicate its size on the diagram.

Analysis

1. Describe the appearance of a pork worm larva.
microscopic, round, like a hot dog, spiral

2. Why might it be difficult to find larva embedded in muscle when meat inspectors use visual checking methods in packing houses to screen for pork worm contamination?
Pork worm larvae are microscopic and cannot be identified by a visual inspection.

3. Suggest what inspectors might do to help detect pork worm larvae.
Samples of muscle tissue can be taken and viewed under a microscope.

Expected Results: Students will not be able to see pork worm larvae without the aid of a microscope. Size of the larvae will be close to 100 mm.

Name Date Class

Chapter 26

INVESTIGATE BioLab

Observing Planarian Regeneration

Preparation

Problem
How can you determine if the flatworm *Dugesia* is capable of regeneration?

Objectives
In this BioLab, you will:
- **Observe** the flatworm, *Dugesia.*
- **Conduct** an experiment to determine if planarians are capable of regeneration.

Materials
planarians
petri dish
springwater
camel hair brush
chilled glass slide
binocular microscope
marking pencil or labels
single-edged razor blade

Safety Precautions
Always wear goggles in the lab. Use extreme caution when cutting with a razor blade. Wash your hands both before and after working with planarians.

Skill Handbook
Use the **Skill Handbook** if you need additional help with this lab.

Procedure

1. Obtain a planarian and place it in a petri dish containing a small amount of springwater. You can pick up a planarian easily with a small camel hair brush.
2. Use a binocular microscope to observe the planarian. Locate the animal's head and tail region and its "eyes." Use diagram **A** on page 735 of your text as a guide
3. Place the animal on a chilled glass slide. This will cause it to stretch out.
4. Place the slide onto the microscope stage. While observing the worm through the microscope, use a single-edged razor to cut the animal in half across the midsection. Use diagram **B** on page 735 of your text as a guide.
5. Remove the head end and place it in a petri dish filled with springwater. Label the dish with the date, your name, and the word "head."
6. Add the tail section to a different petri dish and label it as in step 5, marking this dish "tail."
7. Repeat steps 3–6 with a second flatworm and add the correct pieces to the proper petri dishes.
8. Place the petri dishes in an area designated by your teacher.
9. Prepare a data table that will allow you to record the appearance of your flatworms every other day for two weeks. Include diagrams and the number of days since starting the experiment in your data table.
10. Observe your animals under a binocular microscope and record observations and diagrams in your data table.

Name Date Class

Chapter 26

INVESTIGATE BioLab

Observing Planarian Regeneration, *continued*

Analyze and Conclude

1. **Knowledge** To what phylum do flatworms belong? Are planarians free-living or parasitic? What is your evidence?
   Planarians belong to the phylum Playthelminthes. They are free-living and can be found in ponds and streams living on their own.
2. **Observing** What new part did each original head piece regenerate? What new part did each original tail piece regenerate?
   The head regenerated a new tail, and the tail regenerated a new head.
3. **Observing** Which section, head or tail, regenerated new parts faster?
   The head section regenerated a new tail faster.
4. **Interpreting** Are planarians able to regenerate new parts? Would regeneration be by mitosis or meiosis? Explain.
   Yes, planarians can regenerate new body parts. Regeneration occurs through mitosis; no sexual reproduction or formation of gametes was needed.
5. **Thinking Critically** What might be the advantage for an animal that can grow new body parts through regeneration?
   Answers may vary; there is no need to find a mate, able to replace lost body parts, faster than sexual reproduction, identical genetic makeup as original animal
6. **Thinking Critically** Would the term "clone" be suitable in reference to the newly formed planarians? Explain your answer.
   Yes, clones may be formed asexually by mitosis. Regeneration in planarians is a type of cloning.

Data and Observations: Students will observe the flatworms and will be able to see those structures shown in Figure A. Data table design will vary from student to student. Encourage students to diagram their observations. At the end of two weeks, students will observe that a new head has formed on the original tail section, and a new tail on the original head section.

Name Date Class

MiniLab 27-1 Identifying Mollusks

Comparing and Contrasting

Have you ever taken a walk on the beach and filled your pockets with shells, and as you examined them later, wondered what they were? Use the following dichotomous key to determine the names of the shells.

Procedure

Refer to p. 746 of the text to examine the pictured shells.

1. To use a dichotomous key, begin with a choice from the first pair of descriptions.
2. Follow the instructions for the next choice. Notice that either a scientific name can be found at the end of each description, or directions will tell you to go on to another numbered set of choices.

1A One shell............Gastropods see 2
1B Two shells............Bivalves see 3
2A Flat coil............Sundial shell: *Architectonica nobilis*
2B Thick coil............see 4
3A Shelf inside shell............Common Atlantic slipper: *Crepidula fornicata*
3B No shelf inside shell............see 5
4A Spotted surface............Junonia shell: *Scaphella junonia*
4B Lined surface............Banded tulip shell: *Fasciolaria hunteria*
5A Polished surface............Sunray shell: *Macrocallista mimbosa*
5B Rough surface............Lion's paw shell: *Lyropecten nodosus*

Analysis

1. Why is a dichotomous key used for a variety of organisms?
   A dichotomous key divides a group into smaller and smaller groups until each organism is identified.
2. What shell features were easy to pick out using the key? What features were more difficult?
   easy; 1, 2, 3, 4; more difficult, 5, because it requires more interpretation and closer comparison
3. What general feature was used to identify shells?
   one or two shells

Expected Results: Students will classify the pictured shells using the dichotomous key provided.

Name Date Class

MiniLab 27-2 A Different View of an Earthworm

Interpreting Scientific Diagrams

What does an earthworm look like internally? You could look at it many different ways—from the dorsal or ventral side, along the length of the animal (a longitudinal view), or in cross section through a segment.

Procedure

1. Diagram A on page 750 of your text illustrates a longitudinal dorsal view of the internal organs of an earthworm. Note that the segments are numbered.
2. Use Diagram B on page 750 of your text as a guide to how a cross-section slice appears through segment 9.

Analysis

Make your own cross-section diagrams of segments 8 and 12. Label all the parts shown in your diagrams.

Segment 8 will show: muscles, esophagus, heart, dorsal and ventral blood vessels, and nerve cord. Segment 12 will show all the parts from Segment 8 plus the seminal vesicle and calciferous gland.

Expected Results: Student diagrams will reflect their ability to translate information from a longitudinal view to a cross-sectional view.

Name Date Class

DESIGN YOUR OWN BioLab

How do earthworms respond to their environment?

Chapter 27

PREPARATION

Problem
How do earthworms respond to light, different surfaces, moist and dry environments, and warm and cold environments?

Hypotheses
Place your worm in a tray with some moist soil. Watch your worm for about 5 minutes, and record what you observe. Make a hypothesis based on your observations about what the worm might do under conditions of light and dark, rough and smooth surfaces, moist and dry surfaces, and warm and cold conditions. Limit your investigation as time requires.

Objectives
In this BioLab, you will:

- **Measure** the sensitivity of earthworms to different stimuli, including light, water, and temperature.
- **Interpret** earthworm responses according to terms of adaptations that promote their survival.

Possible Materials

live earthworms	paper towels
glass pan	sandpaper
culture dishes	warm tap water
thermometer	water
dropper	penlight
ice	ruler
black paper	cotton swabs
hand lens or stereomicroscope	

Safety Precautions
Be sure to treat the earthworm in a humane manner at all times. Wet your hands before handling earthworms. Always wear goggles in the lab.

Skill Handbook
Use the **Skill Handbook** if you need additional help with this lab.

PLAN THE EXPERIMENT

1. As a group, make a list of possible ways you might test your hypothesis. Keep the available materials in mind as you plan your procedure.
2. Be sure to design an experiment that will test one variable at a time. Plan to collect quantitative data. Make sure to incorporate a control.
3. Record your procedure and list materials and amounts you will need. Design and construct a data table for recording your findings.

Check the Plan
Discuss the following points with other group members.

1. What data will you collect, and how will they be recorded?
2. Does each test have one variable and a control? What are they?
3. Each test should include measurements of some kind. What are you measuring in each test?
4. How many trials will you run for each test?
5. Assign roles for this investigation.
6. ***Make sure your teacher has approved your experimental plan before you proceed further.***
7. Carry out your experiment. **CAUTION:** ***Return earthworms to the container the teacher has provided.***

Name Date Class

DESIGN YOUR OWN BioLab

How do earthworms respond to their environment?, *continued*

Chapter 27

ANALYZE AND CONCLUDE

1. **Checking Your Hypothesis** Which surface did the worm prefer? Explain.
   a rough surface; the worm moves more easily on a rough surface
2. **Interpreting Observations** In which temperature was the worm most active? Explain.
   an intermediate temperature; an earthworm is ectothermic so its level of activity depends upon the surrounding temperature
3. **Observing and Inferring** How did the earthworm respond to light? Of what survival value is this behavior?
   moved away from light; earthworms are safer from predators in the soil where it is dark
4. **Observing and Inferring** How did the earthworm respond to dry and moist environments? Of what survival value is this behavior?
   preferred a moist environment; earthworm's skin must remain moist or the animal will dry out and die
5. **Drawing Conclusions** Were your hypotheses supported by your data? Why or why not?
   Students who made hypotheses that the worms would prefer moist environments, intermediate temperatures, darkness, and rough surfaces most likely would have their hypotheses supported by data.

Data and Observations: Most likely, earthworms will avoid light and extremes of temperature, move quickly on a rough surface, and prefer a moist surface.

MiniLab 28-1

Comparing and Contrasting

Crayfish Characteristics

There are more species of arthropods than all of the other animal species combined. This phylum includes a variety of adaptations that are not found in other animal phyla.

Procedure

1. Examine a preserved crayfish. **CAUTION:** ***Wear disposable latex gloves and use a forceps when handling preserved material.***
2. Prepare a data table with the following arthropod traits listed: body segmentation, jointed appendages, exoskeleton, sense organs, jaws.
3. Observe the crayfish. Fill in your data table, indicating which of the arthropod traits you observed.
4. Gently lift the edge of the body covering where the legs attach to the body. Look for feathery structures. These are gills and are part of the animal's respiratory system. **CAUTION:** ***Wash hands with soap and water after handling preserved materials.***

Analysis

1. Do crayfish have all of the traits listed above?
   Yes: all traits listed are the major traits that characterize Arthropoda.
2. Make a hypothesis as to how crayfish locate food.
   Student answers may vary: visually through their eyes; detection of odor with antennae; direction of movement through antennae.

Expected Results: All arthropod traits can be observed on the crayfish.

MiniLab 28-2

Comparing and Contrasting

Comparing Patterns of Metamorphosis

Insects undergo a series of developmental changes called metamorphosis. But not all insects follow the same pattern of metamorphosis.

Procedure

1. Use the data table below.
2. Examine the three life stages of a grasshopper. Complete the information called for in your data table. **CAUTION:** ***Wear disposable latex gloves and use forceps to handle preserved insects.***
3. Examine the four life stages of a moth. Complete the information called for in your data table.

Data Table

Insect	Grasshopper			Moth			
Stage	egg	nymph	adult	egg	larva	pupa	adult
Locomotion Method	no	yes	yes	no	yes	no	yes
Feeding Method	no	yes	yes	no	yes	no	yes
Able to Reproduce	no	no	yes	no	no	no	yes

Analysis

1. What are the differences between the stages of metamorphosis of a grasshopper and those of a moth?
   A grasshopper has incomplete metamorphosis. The moth has complete metamorphosis.
2. Correlate the ability to move with ability to feed.
   The stages that were able to move are the only stages in which feeding occurs.
3. Compare a nymph stage with an adult stage.
   The nymph stage looks like a small adult, but it lacks wings and is sexually immature.

Expected Results: See data table above.

Name Date Class

DESIGN YOUR OWN BioLab

Will salt concentration affect brine shrimp hatching?

Chapter 28

PREPARATION

Problem

How can you determine the optimum salt concentration for the hatching of brine shrimp eggs?

Hypothesis

Decide on one hypothesis that you will test. Your hypothesis might be that increased salt concentrations result in an increase in the number of eggs hatched.

Objectives

In this BioLab, you will:

- **Analyze** how salt concentration may affect brine shrimp hatching.
- **Interpret** your experimental findings.

Possible Materials

beakers or plastic bottles
labels or marking pencil
graduated cylinder
brine shrimp eggs
clear plastic trays
salt (noniodized)
balance
water

Safety Precautions

Wear protective eye goggles when preparing solutions.

Skill Handbook

Use the **Skill Handbook** if you need additional help with this lab.

PLAN THE EXPERIMENT

1. Decide on a way to test your group's hypothesis. Keep the available materials in mind as you plan your procedure. Be sure to include a control. For example, you might place brine shrimp eggs in two trays—one with the salt concentration of the water brine shrimp normally inhabit, and one with a different salt concentration.
2. Decide how long you will make observations and how you will judge the extent of egg hatching.
3. Decide on the number of different salt water concentrations to use and what these concentrations will be. Review the steps needed to prepare solutions of different concentrations.

Check the Plan

Discuss the following points with other group members to decide on the final procedure for your experiment.

1. What is your one independent variable? Your dependent variable?
2. What will be your control?
3. How much water will you add to each tray and how will you measure the same number of eggs to be used in each tray?
4. Will it be necessary to control variables such as light and temperature?
5. What data will you collect and how will it be recorded?
6. ***Make sure your teacher has approved your experimental plan before you proceed further.***
7. Carry out your experiment.

Name Date Class

DESIGN YOUR OWN BioLab

Will salt concentration affect brine shrimp hatching?, *continued*

Chapter 28

ANALYZE AND CONCLUDE

1. **Interpreting Data** Using specific numbers from your data, explain how salt concentration affects brine shrimp hatching.
   Student answers may vary; maximum hatching will occur at 1–4% salt concentration.
2. **Drawing a Conclusion** Was your hypothesis supported? Explain.
   Student answers will vary depending on their original hypothesis.
3. **Identifying and Controlling Variables** What were the independent and dependent variables? What were some of the variables that had to be controlled?
   The independent variable was the salt concentration. The dependent variable was the number of eggs hatched. Other variables that had to be controlled were the temperature of the water, amount of water used, and the number of eggs used.
4. **Hypothesizing** Formulate a hypothesis that explains why high salt concentrations may be harmful to brine shrimp hatching.
   Too high a salt concentration may result in water loss from the egg or larva, leading to dessication.
5. **Classifying** Classify brine shrimp. Identify their kingdom, phylum, class, order, family, genus, and species.
   animal, Arthropoda, Crustacea, Anostraca, *Artemia salina*

Data and Observations: Ideal salt concentrations for hatching will be in the 1–4% range. No salt present or concentrations higher than 4% will result in no or few larvae hatching.

Name Date Class

MiniLab 29-1 Examining Pedicellariae

Observing and Inferring

Echinoderms move by tube feet. They also have tiny pincers on their skin called pedicellariae.

Procedure

1. Observe a slide of sea star pedicellariae under low-power magnification. **CAUTION:** ***Use caution when working with a microscope and slides.***
2. Record the general appearance of one pedicellaria. What does it look like?
3. Make a diagram of one pedicellaria under low-power magnification.
4. Indicate the size of one pedicellaria in micrometers.

Analysis

1. Describe the general appearance of one pedicellaria.
Student answers will vary—pincherlike, plierslike, forcepslike

2. What is the function of this structure?
allows animal to grasp objects, clean itself of debris, protection

3. Explain how the structure of pedicellariae assists in their function.
These pincherlike organs allow the sea star to pinch potential predators when they touch the animal. Their shape also allows the sea star to clean itself by picking off materials that become stuck to its body.

Expected Results: Students will observe that pedicellariae have a pincherlike appearance and measure close to 250 µm in length.

Name Date Class

MiniLab 29-2 Examining a Lancelet

Observing

Branchiostoma californiense is a small, sea-dwelling lancelet. At first glance, it appears to be a fish. However, its structural parts and appearance are quite different.

Procedure

1. Place the lancelet onto a glass slide. **CAUTION:** ***Wear disposable latex gloves and handle preserved material with forceps.***
2. Use a dissecting microscope to examine the animal. **CAUTION:** ***Use care when working with a microscope and slides.***
3. Prepare a data table that will allow you to record the following: General body shape, Length in mm, Head region present, Fins and tail present, Nature of body covering, Sense organs such as eyes present, Habitat, Segmented body.
4. Indicate on your data table if the following can easily be observed: gill slits, notochord, dorsal hollow nerve cord.

Analysis

1. How does *Branchiostoma* differ structurally from a fish? How are its general appearance and habitat similar to those of a fish?
It has no sense organs, gills, or fins. But it does have a distinct head area. Fishes are classified in the same phylum as lancelets—Chordata—but in subphylum Vertebrata, whereas lancelets are in subphylum Cephalochordata.

2. Explain why you were not able to see gills, notochord, and a dorsal hollow nerve cord.
These structures are all internal organs.

3. Using its scientific name as a guide, where might the habitat of this species be located?
in the ocean along the coast of California

Expected Results: Lancelets have a long, tubular body shape; length close to 50 mm, head region, tail-like posterior; smooth body; no sense organs. Students will not be able to see a notochord, gill slits, or a dorsal hollow nerve cord.

Name Date Class

INVESTIGATE BioLab

Observing Sea Urchin Gametes and Egg Development

Chapter 29

Preparation

Problem

How can you induce a sea urchin to release its gametes?

Objectives

In this BioLab, you will:

- **Induce** sea urchins to release their gamete cells.
- **Observe** living sperm and egg cells under the microscope.
- **Observe** developmental changes in a fertilized sea urchin egg.

Materials

live sea urchins
sea water
glass slides and coverslips
syringe filled with potassium chloride
beakers
petri dish
dropper
microscope
test tube

Safety Precautions

Always wear goggles in the lab.

Skill Handbook

Use the **Skill Handbook** if you need additional help with this lab.

Procedure

1. Fill a small beaker (250mL) with sea water.
2. Obtain a live sea urchin from your teacher and locate an area of soft tissue next to its mouth.
3. Using a syringe, your teacher will insert the needle into this soft tissue and inject the syringe contents into the sea urchin.
4. Turn your animal so that its mouth is facing up and place it in a petri dish. **CAUTION:** ***Use care in handling live animals.***
5. Wait a minute or two, then check the petri dish. If the sea urchin is male, a milky white mass of sperm will be present in the dish. If it is female, a yellow orange mass of eggs will be seen.
6. If you have a female sea urchin, hold her upside down directly over the seawater-filled beaker and allow the eggs to fall directly into the water.
7. If your urchin is male, use a dropper to add several drops of sperm from the petri dish to your beaker of sea water.
8. Check with your classmates to see who has a male and who has a female sea urchin. Share gamete cells.
9. Use a clean dropper to transfer a drop of sperm from the beaker to a microscope slide. Observe under low power without a coverslip.
10. Add a coverslip and observe under high power. Note the movement of sperm. Draw several sperm cells and indicate their size in µm. Note the approximate number of sperm cells present.
11. Repeat steps 9 and 10 for egg cells. In step 10, use only low power to observe egg cells.
12. For this step, work with a partner. While one partner transfers some sperm to the slide with egg cells using a clean dropper, the other partner should observe under low power.
13. Observe the process of fertilization and note any changes that occur to the egg. Record your observations in a data table.

Name Date Class

INVESTIGATE BioLab

Observing Sea Urchin Gametes and Egg Development, *continued*

Chapter 29

14. When fertilization has been accomplished, place the fertilized eggs in a test tube filled with 10 mL of seawater. Label your tube and observe the eggs 24 hours later under low power. Record any changes that you see. **CAUTION:** ***Wash your hands immediately after working with animals.***

Analyze and Conclude

1. **Compare and Contrast** Compare eggs and sperm, noting numbers released, numbers observed under low power, size, and ability to move.
   Sperm are motile, thousands are present on a slide, the size is close to 10 µm; eggs are not motile, less than 100 on a slide, size is close to 150 µm.
2. **Predicting** Based on the pattern of fertilization, predict the reason for the large number of gametes released in nature.
   Student answers may vary; fertilization is external, so chances of a sperm meeting an egg are less; a large number of gametes released improves the chances of fertilization.
3. **Observing** Describe the behavior of sperm when they first come in contact with an egg.
   Sperm tend to cluster around the outer edge of the egg.
4. **Observing** How does an unfertilized egg differ in appearance from a fertilized egg? Draw both eggs in your data table.
   A fertilized egg forms a clear membrane around the outside edge of the egg shortly after fertilization.

Data and Observations: Students will observe the differences between egg and sperm. They will determine that developmental changes occur within minutes after fertilization.

MiniLab 30-1 Measuring Breathing Rate in Fishes

Experimenting

Fishes are able to extract oxygen from water as it flows over their gills. Their rate of breathing is related to the availability of oxygen in the water. More oxygen results in a slower breathing rate. The breathing rate of a fish can be estimated by counting the number of times per minute its gill covers open to allow water to flow across its gills.

Procedure

1. Make a hypothesis about how the breathing rate of a fish may be influenced by a change in temperature of the water in which it is swimming. Record your hypothesis.
2. Fill a beaker with non-chlorinated water. Let the beaker sit until the water reaches room temperature (about 20°C). Measure the temperature with a thermometer and record it in a data table.
3. Add a small goldfish. **CAUTION:** ***Handle animals with care.*** Wait 5 minutes for the fish to acclimatize to the water temperature.
4. Count the number of times the goldfish's gill covers open in one minute. This is a measure of the rate at which the fish is breathing. Record your result in a data table.
5. Repeat step 4 four more times. Find an average for your trials and record these data in your table.
6. Remove the goldfish from the beaker. Add one ice cube made from non-chlorinated water to the beaker.
7. When the ice cube has melted, record the water's temperature.
8. Repeat steps 3–5. Remove goldfish from the beaker, and add another ice cube to the beaker. Repeat step 7.
9. Repeat steps 3–5. **CAUTION:** ***Wash your hands after working with animals.***

Analysis

1. How does water temperature influence the rate at which breathing occurs in a goldfish?
   The breathing rate decreases as the water temperature decreases.
2. Do your data support your hypothesis? Explain.
   Student answers will vary. Most students will have hypothesized that a decrease in water temperature results in an increase in breathing rate.

3. Was the breathing rate faster or slower in colder water compared with warmer water?
   slower in cold water
4. How might the amount of oxygen in cold water compare with that in warm water? Explain your answer.
   There must be more dissolved oxygen in cold water as compared with warm water because the fishes didn't need to breathe as often in the colder water.
5. Sequence the events associated with a fish obtaining oxygen. Start with a molecule of oxygen in water and be sure to include the capillaries located in the fish's gills.
   An oxygen molecule in water passes over the gills, diffuses into gill capillaries, passes into the bloodstream, and is pumped into body cells by heart.

Expected Results: Breathing rate will decrease as water temperature decreases.

Name Date Class

MiniLab 30-2

Comparing and Contrasting

Looking at Frog and Tadpole Adaptations

An adult frog and its larval stage are adapted to different habitats. How are the structures of a frog and a tadpole adapted to their environments?

Procedure

1. Use the data table below.
2. Examine a living or preserved adult frog and larval (tadpole) stage. **CAUTION:** ***Wear disposable latex gloves and use a forceps when handling preserved specimens.***
3. Observe the first seven traits listed. Complete your data table for these observations.
4. Use references to fill in the information for the last three traits listed.

Data Table

Trait of information	Tadpole	Adult
Limbs present?		
Eyes present?		
Tympanic membrane present?		
Tail present?		
Mouth present?		
Nature of skin (color and texture)		
General size		
Respiratory organ type		
Diet		
Habitat		

Analysis

1. Explain how hind leg musculature aids in adult frog survival.
   allows frog to escape predators, aids in swimming
2. Correlate the type of respiratory organ in an adult and a tadpole with their differing habitats.
   Diffusion of oxygen occurs between lungs and air, and between gills and water.

Name Date Class

MiniLab 30-2

Looking at Frog and Tadpole Adaptations, *continued*

3. Correlate the type of appendages (arm, leg, tail) in an adult and a tadpole with their differing habitats.
   A tadpole has a tail for swimming in water. An adult frog has legs for jumping away from predators and catching prey.
4. Correlate mouth size in an adult and a tadpole with their differing diet.
   A tadpole has a small mouth for eating algae. An adult frog has a large mouth for eating insects, worms, and fishes.
5. Explain how eyes may aid in the survival of both stages.
   seeing food and predators
6. Explain why the tympanic membrane may not be essential to the survival of a tadpole.
   An adult can hear mating calls to enable reproduction. Tadpoles do not reproduce.
7. Predict how skin color and texture aids in adult frog survival. **CAUTION:** ***Wash your hands after working with live or preserved animals.***
   protective coloration from predators, thin skin aids in gas exchange

Expected Results: Tadpoles have a tail, gills, no tympanic membrane, and no limbs at early stages. Adults have lungs. All other traits are shared. Coloration of stages may vary with species being observed. Tadpoles feed on algae, adults on insects, worms, and small fishes. Tadpoles are confined to aquatic habitats; adults live both on land and in water.

Name Date Class

INVESTIGATE BioLab

Development of Frog Eggs

Chapter 30

Preparation

Problem
How does temperature affect the development of frog eggs?

Objectives
In this BioLab, you will:
- **Compare** development of frog eggs at varying temperatures.
- **Distinguish** among various stages of development.

Materials
Ringer's solution
4 culture dishes
lightbulbs with source of electricity
thermometer
binocular microscope
frog eggs, *Xenopus laevis*
flashlight

Safety Precautions
Always wear goggles in the lab. Wash hands before and after each observation.

Skill Handbook
Use the **Skill Handbook** if you need additional help with this lab.

Procedure

1. Obtain four culture dishes of Ringer's solution and fertilized frog eggs.
2. Make a data table similar to the one shown for sketching of stages of development of the eggs.
3. Observe your eggs and determine their stage of development. At room temperature, you should see the two-cell stage about 1.5 hours after fertilization, the eight-cell stage at 2.25 hours, the 32-cell stage at 3 hours, the late gastrula stage at 9 hours, and a visible head area between 18 and 20 hours. Hatching will occur at about 50 hours (about two days).
4. Set up the appropriate numbers of lightbulbs of different wattages over the water to keep the temperatures at 20°, 25°, and 30°C.
5. Place a dish of eggs in the refrigerator. Measure the temperature of your refrigerator. Keep a flashlight on in the refrigerator at all times so that you have only one variable, the temperature.
6. Make a hypothesis about how temperature will affect development of the eggs.
7. Set up a time schedule for observations and making sketches based on what stage you are observing. Make observations until the eggs hatch. Record your observations in a journal.
8. Observe your eggs under the microscope according to the schedule you have made. Draw sketches of your eggs in the data table.

Data Table

Temperature	Day 1	Day 2	Day 3
30° C			
25° C			
20° C			
refrigerator			

Name Date Class

INVESTIGATE BioLab

Development of Frog Eggs, *continued*

Chapter 30

Analyze and Conclude

1. **Interpreting Observations** Which eggs develop the fastest? The slowest? Explain.
Egg development and cell division occur more quickly at warmer temperatures. Cell division is slower at lower temperatures.
2. **Interpreting Observations** Did your data support your hypothesis? Explain.
Check students' data to see how they supported their hypotheses.
3. **Drawing Conclusions** What advantage is it for frogs to have eggs that develop at different rates that correspond to different temperatures?
The tadpoles will hatch when temperature is most conducive for survival. Perhaps more food is available at warmer temperatures.
4. **Thinking Critically** What would happen if frog eggs developed when the weather was still cold in the spring?
The tadpoles might not survive or adequate food may not be available when the water is cold.

Data and Observations: Eggs develop more quickly at warmer temperatures and more slowly at cooler temperatures.

Name Date Class

MiniLab 31-1

Comparing and Contrasting

Comparing Feathers

Birds have two kinds of feathers. Contour feathers used for flight are found on a bird's body, wings, and tail. Down feathers lie under the contour feathers and insulate the body.

Procedure

1. Examine a contour feather with a hand lens, and make a sketch of how the feather filaments are hooked together.
2. Examine a down feather with a hand lens. Draw a diagram of the filaments of the down feather.
3. Fan your face with each feather separately. Note how much air is moved past your face by each type of feather. **CAUTION:** ***Wash your hands after handling animal material.***

Analysis

1. How does the structure of a contour feather help a bird fly?
   Contour feathers are sleek and streamlined and can move a lot of air.
2. How does the structure of a down feather keep a bird warm?
   Down feathers are irregular in shape and, when piled together, form air spaces that trap body heat.
3. How can you explain the differences you felt when fanning with each feather?
   The contour feather barbules stay together and move air. The down feather barbules cannot move air because they do not stay together.

Expected Results: Contour feathers consist of barbules with hooks that connect the barbs, forming a streamlined feather. Down feathers do not have hooks on their barbules. They are soft and do not take on a specific shape. More air can be moved with the contour feather.

Name Date Class

MiniLab 31-2

Comparing and Contrasting

Feeding the Birds

In the winter, it may be difficult for some birds to find food, especially if you live in an environment often blanketed with snow. Making a bird feeder and watching birds feed can be an enjoyable activity for you that may save some birds from starvation. If you do begin feeding birds in the winter, continue to feed them until natural food again becomes available in the spring.

Procedure

1. Obtain several large, plastic milk bottles. Cut two holes 5 cm from the base on opposite sides of each bottle, each about 8 cm^2. These are the openings birds will use to find the food inside.
2. Place small drainage holes in the bottom of each bottle. Hang the bottles from wires strung through small holes in the neck of each one.
3. Place a different kind of seed (sunflower seeds, hulled oats, cracked corn, wheat, thistle, millet) in different bottles. Add new seed when needed.
4. Using a bird guide, make a list of numbers and kinds of birds that frequent each feeder, noting the type of food offered.

Analysis

1. What type of seed attracted the largest variety of bird types?
   Sunflower seeds will attract a large variety of birds.
2. Did any birds visit more than one feeder?
   It is likely that birds will visit the same feeder over and over again, unless the feeder runs out of food.
3. What do you think an ideal bird food would be?
   Ideal bird food would have a mixture of a variety of seeds that would appeal to many different birds.

Expected Results: Depending on region, students may find cardinals, jays, woodpeckers, nuthatches, juncos, chickadees, sparrows, finches, tufted titmouse, mourning doves, and many other local and migrating bird species.

Name Date Class

DESIGN YOUR OWN BioLab

Which egg shape is best?

Chapter 31

PREPARATION

Problem

What shape would be best for an egg to reduce the distance it could roll if pushed from a nest?

Hypotheses

There are several hypotheses that you can test. Your hypothesis might be that egg shape influences the distance an egg rolls, or that shape determines the tightness of circular rolling patterns.

Objectives

In this Biolab, you will:

- **Design** an experiment to test your hypotheses.
- **Model** different egg shapes and egg masses.
- **Experiment** to test your hypotheses.
- **Draw conclusions** based on your experimental data.

Possible Materials

clay
cardboard ramp
ruler
string
hard-boiled egg
Ping-Pong ball
golf ball
balance
protractor

Safety Precautions

Always wear goggles in the lab.

Skill Handbook

Use the **Skill Handbook** if you need additional help with this lab.

PLAN THE EXPERIMENT

1. Decide on a way to test your group's hypothesis. Keep the list of available materials in mind as you plan your procedure.
2. There are a number of questions to be asked before starting. Here are some suggestions. How will you incorporate a control? How many egg shapes will you test? How will you model your egg shapes? How many trials will you perform? How might you keep egg models identical in mass? How will you measure the angle of the cardboard ramp? Where will you start to measure distance rolled?

Check the Plan

Discuss the following points with other group members to decide the final procedure for each of your experiments.

1. What is your independent and dependent variable?
2. How will you eliminate all other variables?
3. What data will you collect? How many trials will you run?
4. Will you need a data table and how might it be organized?
5. ***Make sure your teacher has approved your experimental plan before you proceed further.***
6. Carry out your experiments.

Name Date Class

DESIGN YOUR OWN BioLab

Which egg shape is best?, *continued*

Chapter 31

ANALYZE AND CONCLUDE

1. **Hypothesizing** Record your hypothesis.
   Student answers will vary. Example: Egg shape will not influence path that the egg follows.
2. **Interpreting Data** Describe your results after testing your hypothesis.
   Student answers will vary. Example: Round eggs roll in straight lines, whereas oval or pointed eggs roll in a circular pattern.
3. **Concluding** Do your data support your hypothesis? Explain using both quantitative and qualitative observations.
   Student answers will vary. Example: Heavy round eggs rolled farther (26 cm) than lighter round eggs (22 cm).
4. **Identifying Variables** What were your independent and dependent variables?
   Student answers will vary. Example: independent variable, egg shape; dependent variable, distance rolled.
5. **Concluding** In general, how does mass influence the distance an egg will roll? How does egg shape influence the distance an egg will roll or the pattern taken when it rolls?
   A greater mass results in a longer distance rolled. Oval or pointed eggs roll a shorter distance and form a circular path; the more pointed the end, the tighter the circular path.
6. **Predicting** Predict why egg shape or mass may be helpful adaptations when considering the variety of habitats where birds live.
   Round shapes aid birds that nest on flat ground. Oval or pointed eggs aid birds that nest on slanted ground or on cliffs.

Data and Observations: Round eggs will roll the farthest. Eggs of higher mass will roll farther than lighter eggs. Pointed eggs or oval (normal egg shape) eggs will roll in a circular pattern, and total distance will be less than total distance rolled for round eggs.

Name Date Class

MiniLab 32-1 Anatomy of a Tooth

Observing

Most mammals have teeth. Teeth typically have evolved into a variety of shapes, depending on the diet of the animal.

Procedure

1. Examine a prepared slide of a human tooth under low-power magnification. The slide is a longitudinal section view of the tooth. **CAUTION:** ***Use caution when working with a microscope and slides.***
2. Locate the different areas that form the human tooth. From outside to inside they are: enamel, dentine, and pulp cavity. Each area varies in thickness and texture. You will have to move the slide from left to right or up and down to see all areas.
3. Diagram the appearance of the entire cross section as it appears under low power, labeling the three areas.

Analysis

1. Which area is the thickest? Thinnest?
   Dentine is thickest; enamel or pulp is thinnest.
2. Describe the nature of the three areas. Use such terms as compact, loose, bonelike, vascular (with blood vessels).
   Enamel is compact and bonelike; dentine is also compact and bonelike; pulp is vascular and looser.
3. Which area of the three would you expect to be the toughest? Explain why.
   Enamel; the outermost layer receives the most wear.
4. Draw cross sections of the two tooth shapes shown on page 869 of your text, at the points marked x and y on each diagram. Include the three areas just studied and label each area.
   Examine student drawings to make sure they have located each layer appropriately.

Expected Results: Enamel is the outermost, most compact and bonelike thinnest layer. Dentine is the thickest layer, also compact and bonelike. Pulp is the center layer and appears vascular and the least dense of the three layers.

Name Date Class

MiniLab 32-2 Mammal Skeletons

Observing

Owls are predators. They feed on small mammals as well as on birds. After eating a meal, the tough indigestible parts of prey, such as bones, are regurgitated as pellets. The skeletons of a variety of small mammals can be studied by examining owl pellets.

Procedure

1. Place an owl pellet onto a sheet of paper toweling.
2. Use a forceps to remove a very small amount of the outer covering.
3. Prepare a wet mount of this material and observe under low-, then high-power magnification. Diagram what you see. Use forceps to open the pellet and remove all bones that are present. Look especially for skulls. (Note: Skulls may be small and certain parts, such as lower jaws, will be separated.)
4. Identify each mammalian skull using Diagrams A and B on page 871 of your text as a guide.
5. Attempt to reconstruct the skeleton for an entire animal. You may wish to glue the skeletal piece onto a piece of cardboard. **CAUTION:** ***Wash your hands after handling animal materials.***

Expected Results: Hair comprises the outer covering of the pellet. Students will be able to differentiate between vole and shrew skulls.

Analysis

1. What was the outer covering on the pellet? What does this tell you about the contents of the pellet? Explain.
   Hair; the prey must be a mammal if hair is present.
2. How many vole and shrew skulls were present in your pellet?
   Answers may vary; there may be 1–4 skeletons per pellet.
3. How were you able to differentiate the skulls of these two mammals?
   by size and dentition patterns
4. Predict if voles and shrews are herbivores or carnivores based on appearance of their teeth. Explain.
   Both are herbivores because they lack canines or incisors for capturing or tearing prey.

Name Date Class

INTERNET BioLab

Domestic Dogs Wanted

Chapter 32

Preparation

Problem
What are the most popular breeds of dog?

Objectives
In this BioLab, you will:
- **Observe** the characteristics of dogs.
- **Record** the popularity of different dog breeds.
- **Use the Internet** to collect and compare data from other students.
- **Compare** the most popular dog breeds with the most common breeds found in animal shelters.

Materials
computer access to the Internet

Skill Handbook
Use the **Skill Handbook** if you need additional help with this lab.

Procedure

1. Use both data tables.
2. Go to the Glencoe Science Web Site at www.glencoe.com/sec/science to find links to sites that rank the most popular breeds of dog. Also find links to animal shelters and pet adoption agencies.
3. From your data, determine the five most popular dog breeds. Record your findings in Data Table I.
4. Find pictures of these dogs and record the physical characteristics unique to their breeds in Data Table I.
5. At the Glencoe Science Web Site find links to five web sites of animal shelters and pet adoption agencies across the country. Many of these sites post pictures and other information about the dogs that are up for adoption.
6. From your data, determine which breeds are most commonly found in animal shelters. Record the five most common breeds in Data Table II.
7. Find pictures of these dog breeds and record their unique characteristics in Data Table II.

Data Table I

Most Popular Dog Breeds		
Rank	Breed	General Characteristics

Name Date Class

INTERNET BioLab

Domestic Dogs Wanted, *continued*

Chapter 32

Data Table II

Most Common Dog Breeds Found in Animal Shelters		
Rank	Breed	General Characteristics

Analyze and Conclude

1. **Using the Internet** Compare the data in Tables I and II. Are there similarities? How might you explain any similarities or differences? Consider the time of year, the age of the dog, and any breeds of dog that are popularized by television or other advertising while formulating your answer.
When students come across a mutt or a mixed breed on an Internet site, it may list the dominant breed only. Have them make observations based on the dominant breed.
2. **Thinking Critically** Are there any breeds that are popular as pets that are commonly found in animal shelters? Propose an explanation for this.
Some large dogs, such as German shepherds and golden retrievers, are often bought as puppies but abandoned or given up when they grow too large.
3. **Comparing and Contrasting** What characteristics do domestic dogs have that are the same as all other carnivores? As all other mammals?
carnivores: canine teeth, claws, strong jaws; mammals: hair, warm-blooded, teeth
4. **Problem Solving** Propose a local program to reduce the number of stray and abandoned dogs.
Many programs of this sort can run through a local humane society or "no-kill" animal shelter. Often these programs involve educating the public on spaying or neutering their animals to reduce unwanted litters.

Data and Observations: Make sure students are sampling data from different animal shelters around the country. Many shelters will list only the breed of dog and possibly a picture. Ask students to infer the different dog characteristics according to the breed.

Name Date Class

MiniLab 33-1 Testing an Isopod's Response to Light

Experimenting

Isopods, the pill bugs and sow bugs, are common arthropods on sidewalks or patios. They are actually land crustaceans and respire through gill-like organs that must be kept moist at all times.

Procedure

1. Use the data table on the right.
2. Prepare a plastic dish using the diagram on page 890 of your text as a guide. Moisten the paper toweling.
3. Place six isopods in the center of the dish and add the cover. Place the dish near a lamp or next to a classroom window with light. Have the light strike the dish as shown in the diagram. **CAUTION:** ***Treat isopods gently.***
4. Wait five minutes and observe the dish. Count and record in your data table the number of isopods on the dark or light side. This is your "five minute observation."
5. Repeat step 4 three more times, waiting five minutes before each observation.

Data Table

Observation in minutes	Number of isopods present	
	Light side	Dark side
5		
10		
15		
20		
25		

Expected Results: Isopods will move toward low light.

Analysis

1. Do isopods tend to move toward light or dark areas? Support your answer with specific numbers from your data.
   dark areas; majority of isopods were found under dark paper side
2. Might the behavior of isopods toward light or darkness be innate or learned? Explain your answer.
   Explanations may vary, but the behavior is innate. This type of behavior would not have been learned during the experiment because isopods have this behavior as soon as they hatch and it may not be under their conscious control.
3. What might be the adaptive advantage for the observed isopod behavior and their response to light? Explain how natural selection may have influenced this isopod behavior.
   Isopods must remain moist to survive. Dark areas in nature are more likely to be moist than light areas. Through natural selection, those isopods that innately moved toward dark areas survived and passed this genetic trait on to their offspring.
4. Prepare a bar graph that depicts your data.
   Students' bar graphs should depict the data they collected.

Name Date Class

MiniLab 33-2 Solving a Puzzle

Experimenting

You are given a bunch of keys and asked to open a door. How do you go about finding the right key? Several attempts are needed and then finally the door opens. The next time you are asked to perform the same task, can you go directly to the correct key? Chances are, you can. You have learned how to solve this problem.

Procedure

1. Use the data table on the right.
2. Obtain a paper puzzle from your teacher.
3. Time how long it takes you to assemble the puzzle pieces into a perfect square.
4. Record the time it took and call this trial 1.
5. Disassemble the square and mix the pieces.
6. Repeat step 3 for four more trials.

Data Table

Trial	Time needed to complete square puzzle
1	
2	
3	
4	
5	

Expected Results: Student times needed to complete the puzzle will decline with each trial.

Analysis

1. Using your data, explain how the time needed to complete the puzzle changed from trial 1 to trial 5.
   times decreased
2. Was the final completion of the puzzle an example of innate behavior? Explain your answer.
   No, the ability to work the puzzle was an example of learned behavior.
3. Was the final completion of the puzzle an example of learned behavior? Explain your answer.
   Yes, because the time it took to do the puzzle decreased as learning took place.
4. Analyze your behavior when solving the puzzle as to the role that imprinting, trial and error, conditioning, and insight may have played in improving your trial times.
   Student answers may vary. Imprinting and conditioning were not involved in this type of learning. Trial-and-error learning was operating with the initial solving of the puzzle, but later trials may have relied upon insight.

Name Date Class

Chapter 33

INVESTIGATE BioLab

Behavior of a Snail

Preparation

Problem

How can you test the behavior of snails to touch stimuli?

Objectives

In this BioLab, you will:

- **Test** the response of snails to touch.
- **Measure** the time needed for habituation to occur after repeated touch stimuli.

Materials

snails, small dish, dropper, scissors, springwater, dissecting microscope, probe constructed from tape, rubber band, and pencil

Safety Precautions

Always wear goggles in the lab. Wash your hands both before and after handling any animals. Use caution when working with live animals.

Skill Handbook

Use the **Skill Handbook** if you need additional help with this lab.

Procedure

1. Use the data table
2. Prepare a stimulator probe by taping a small piece of a cut rubber band to the tip of a pencil.
3. Cover the bottom of a small dish with springwater.
4. Obtain a snail from your teacher and place it in the dish.
5. Use a dissecting microscope to examine and locate its head. Its head has two antennae that it can extend and retract.
6. Place the dish on your desk.
7. Lightly touch the snail's anterior end using the end of the rubber band probe. Note if it responds (yes or no), and record the direction the snail moves. Consider this trial 1.
8. Repeat step 7 for four more trials.
9. Lightly touch the snail's posterior end using the rubber band probe. Record your observations and conduct a total of five trials.

Data Table

Body area	Response to touch		
Trial	Anterior	Posterior	Middle
1			
2			
3			
4			
5			
Habituation studies			
Rate of stimulation	Number of stimulations needed to reach habituation		
	Data and Observations: Both ends are sensitive to touch. Students may detect no response when performing a specific trial.		

Habituation will usually not occur.

10. Repeat step 9, touching the middle of the snail's body.
11. Test the snail's ability to become habituated from stimulation to its anterior end.
    a. Continue to touch the snail's anterior end with the probe every 10 seconds until habituation occurs. Continue testing for a reasonable length of time if habituation does not occur.
    b. Count and record the number of stimulations needed for habituation.

Analyze and Conclude

1. **Hypothesizing** Are the responses shown by snails to touch learned or innate? Explain your answer.
   Student answers may vary. These responses are innate because the animals lack a large enough brain to learn behavior; they do not have the opportunity to learn correct behavior because the first time they do not respond correctly, they may be eaten by a predator.
2. **Observing** Describe the direction that a snail moves when its anterior and posterior ends are stimulated. Does one end appear to be more sensitive than the other? Is the middle sensitive to touch? Is the speed of response slow or rapid?
   The snail moves backward when the anterior end is touched. The snail moves forward when the posterior end is touched. The anterior appears to be more sensitive to touch. The middle is less sensitive to touch.
3. **Hypothesizing** Explain how the behavior of responding to touch may be an adaptation for survival.
   The snail will move away from predators rapidly.
4. **Experimenting** Why did you perform several trails for each experiment involving stimulation of the anterior, posterior, and middle of the snail?
   Student answers will vary; good experimental technique, failure to make contact with probe, etc.
5. **Defining Operationally** Define the term habituation.
   Habituation is the lack of a response after repeated stimulation.
6. **Concluding** Explain how your data may be used to support the observation that snails are not easily habituated to touch. Use actual data to support your answer.
   Snails continued to respond after several minutes of stimulation to their anterior ends.
7. **Predicting** How might this lack of habituation serve as an adaptation for survival?
   Continued stimulation from a predator would bring about continued response of moving away from that predator.

Name Date Class

MiniLab 34-1 Examine Your Fingerprints

Comparing

Fingerprints play a major role in any police investigation. Because a fingerprint is an individual characteristic, extensive FBI fingerprint files are used for identification in criminal cases.

Procedure

1. Press your thumb lightly on the surface of an ink pad.
2. Roll your thumb from left to right across the corner of an index card, then immediately lift your thumb straight up from the paper.
3. Repeat the steps above for your other four fingers, placing the prints in order across the card.
4. Examine your fingerprints with a magnifying lens, identifying the patterns in each by comparing them with the diagrams on page 925 of your text.
5. Compare your fingerprints with those of your classmates.

Analysis

1. Are the fingerprint patterns on your five fingers all the same?
Although each finger has a unique fingerprint, the general patterns may be the same on different fingers
2. Do any of your fingerprints show the same patterns as those of a classmate?
It is possible that students may share patterns, but not identical fingerprints.
3. Why is a fingerprint a good way to identify a person?
It is unique for each person.

Expected Results: Each finger has a unique fingerprint and each individual has a unique set of fingerprints.

Name Date Class

MiniLab 34-2 Look at Muscle Contraction

Interpreting

Muscle fibers are composed of a number of small functional units called sarcomeres. Sarcomeres, in turn, are composed of protein filaments called actin and myosin. The sliding action of these filaments in relation to each other results in muscle contraction.

Procedure

1. Look at Diagrams A and B on page 937 of your text. Diagram A shows a sarcomere in a relaxed muscle. Diagram B shows a sarcomere in a flexed muscle.
2. Using a centimeter ruler, measure and record the length of a(n): sarcomere in Diagram A, myosin filament in Diagram A, actin filament in Diagram A. Record your data in a table.
3. Repeat step 2 for Diagram B.

Analysis

1. When a muscle contracts, do actin or myosin filaments shorten? Use specific data from your model to support your answer.
Students will learn that actin and myosin do not change in length.
2. How does the sarcomere shorten when parts that make it up don't shorten?
Actin filaments slide over myosin filaments, shortening the sarcomere.

Name Date Class

Design Your Own BioLab

Does fatigue affect the ability to perform an exercise?

Chapter 34

Preparation

Problem
How does fatigue affect the number of repetitions of an exercise you can accomplish? How do different amounts of resistance affect rate of fatigue?

Hypotheses
Hypothesize whether or not muscle fatigue has any effect on the amount of exercise muscles can accomplish. Consider whether fatigue occurs within minutes or hours.

Objectives
In this BioLab, you will:
- **Hypothesize** whether or not muscle fatigue affects the amount of exercise muscles can accomplish.
- **Measure** the amount of exercise done by a group of muscles.
- **Make a graph** to show the amount of exercise done by a group of muscles.

Possible Materials
stopwatch or clock with second hand
graph paper
small weights
wooden box or step stool

Skill Handbook
Use the **Skill Handbook** if you need additional help with this lab.

Plan the Experiment

1. Design a repetitive exercise for a particular group of muscles. Make sure you can count single repetitions of the exercise (for example, one sit-up or one jumping jack) over time.
2. Work in pairs, with one member of the team being a timekeeper and the other member performing the exercise.
3. Consider setting up your experiment so that the amount of resistance is the independent variable. Compare your design with those of other groups.

Check the Plan
1. Be sure the exercises are ones that can be done rapidly and cause a minimum of disruption to other groups in the classroom.
2. Consider how long you will do the activity and how often you will record measurements.

Data Table

Time interval	Number of repetitions
First minute	
Second minute	
Third minute	
Fourth minute	
Fifth minute	

3. ***Make sure your teacher has approved your experimental plan before you proceed further.***
4. Use the data table above to record the number of exercise repetitions per time interval.
5. Carry out the experiment.
6. On a piece of graph paper, plot the number of repetitions on the vertical axis and the time intervals on the horizontal axis.

Name Date Class

Design Your Own BioLab

Does fatigue affect the ability to perform an exercise?, *continued*

Chapter 34

Analyze and Conclude

1. **Making Inferences** What effect did repeating the exercise over time have on the muscle group?
The muscle groups became fatigued and the number of repetitions of the activity went down as more trials were run.
2. **Comparing and Contrasting** As you repeated the exercise over time, how did your muscles feel?
The muscles felt tired and may even have begun to hurt toward the end. It became harder to do the activity, and the strength of each contraction was reduced.
3. **Recognizing Cause and Effect** What physiological factors are responsible for fatigue?
Cells are running out of oxygen and accumulating toxic products such as lactic acid and carbon dioxide as the muscles change to anaerobic processes.
4. **Thinking Critically** How well do you think your fatigued muscles would work after 30 minutes of rest? Explain your answer.
Fatigued muscles should work as well after a rest of 30 minutes as they did before fatigue became a factor. The accumulated lactic acid has been broken down as the oxygen supply in the muscle cells has been replenished during the rest.

Data and Observations: Student graphs should show that the number of exercise repetitions goes down over time as the muscles become tired.

Name Date Class

MiniLab 35-1 Evaluate a Bowl of Soup

Interpreting the Data

As a consumer, you are bombarded by advertising that promotes the nutritional benefits of specific food products. Choosing a food to eat on the basis of such ads may not make nutritional sense. By examining the ingredients of processed foods, you can learn important things about their nutritional content.

Table 35.3

Percentage of Daily Value (DV)	
Carbohydrates	60%
Fat	30%
Saturated Fats	10%
Cholesterol	1.5%
Protein	10%
Total Calories	2000

Procedure

1. Examine the information in the table listing the daily value (DV) of various nutrients. DV expresses what percent of Calories should come from certain nutrients. For instance, in the proposed diet of 2000 Calories, 60 percent of the Calories should come from carbohydrates.
2. Examine the nutritional information on the soup can label on page 957 of your text and compare it with the DV table.

Analysis

1. Does your bowl of soup provide more than 30 percent of any of the daily nutrients? Which ones?
   Yes, sodium.
2. Evaluate the percentage of Calories in soup that are provided by saturated fat.
   A serving of soup contains about 39% of its Calories from saturated fat.
   (6 g/serving × 9 Cal/g = 54 Calories of saturated fat per serving. 54 ÷ 140 = 39%)
3. Is soup a nutritious meal? Explain your answer.
   Basically yes, although the sodium content is high.

Name Date Class

MiniLab 35-2 Compare Thyroid and Parathyroid Tissue

Observing

Although their names seem somewhat similar, the thyroid and parathyroid glands perform rather different functions within the body.

Procedure

1. Use the data table on the right.
2. Use low-power magnification to examine a prepared slide of thyroid and parathyroid endocrine gland tissue. Note: Both tissues appear on the same slide. **CAUTION:** ***Use caution when working with a microscope and prepared slides.***
3. The image on page 964 of your text is a photograph of thyroid and parathyroid tissue. Use it as a guide in locating the two types of endocrine gland tissue under low power and in answering certain analysis questions.
4. Now locate each type of gland tissue under high-power magnification. Draw what you see in the table above. Then use what you learned in the chapter to identify the names of the hormones produced by each gland.

Data Table

Tissue	Drawing	Name of hormone(s) produced
Thyroid		
Parathyroid		

Expected Results: Students will be able to differentiate between thyroid and parathyroid tissue.

Analysis

1. Compare the microscopic appearance of parathyroid tissue to that of thyroid tissue.
   Students may notice that thyroid tissue contains many rather large spaces surrounded by a thin band of cellular tissue, while parathyroid tissue is composed of compact cells with no large spaces or follicles.
2. a. Which tissue type contains follicles (large liquid storage areas)?
   thyroid
   b. What may be present within the follicles?
   stored hormones (thyroxine or calcitonin)
   c. Are follicles composed of cells? Could they produce the hormone associated with this gland tissue? Explain your answer.
   No. No, cells, not storage areas, would be needed to produce the hormone.
   d. Hypothesize what the function may be for the thin layer of tissue that surrounds each follicle.
   This tissue makes the hormones.
3. How might you explain the fact that both thyroid and parathyroid tissue can be seen on the same slide?
   Both glands are located in the same general area of the neck. The parathyroids lie on the thyroid gland itself.

Name Date Class

INVESTIGATE BioLab Average Growth Rate in Humans

Chapter 35

Preparation

Problem
Is average growth rate the same in males and females?

Objectives
In this BioLab, you will:

- **Graph** the average growth rates in males and females.
- **Identify** any differences in the average growth rates of males and females.

Materials
blue pencils
graph paper
red pencils
ruler

Skill Handbook
Use the **Skill Handbook** if you need additional help with this lab.

Procedure

1. Construct a graph for the growth rate data that shows mass on the vertical axis and age on the horizontal axis.
2. On the graph, plot the data shown in the table for the average female growth in mass from ages 8 to 18. Use a ruler to connect the data points with a straight red line.
3. On the same graph, plot the data for the average male growth in mass from ages 8 to 18. Connect these data points with a straight blue line.
4. Construct a second graph that shows height on the vertical axis and age on the horizontal axis.
5. Plot the data for the average female growth in height from ages 8 to 18. Connect the data points with a straight red line.
6. Plot the data for the average male growth in height from ages 8 to 18. Connect these data points with a straight blue line.

Data Table: Averages for growth in humans

	Mass (kg)		Height (cm)	
Age	Female	Male	Female	Male
8	25	25	123	124
9	28	28	129	130
10	31	31	135	135
11	35	37	140	140
12	40	38	147	145
13	47	43	155	152
14	50	50	159	161
15	54	57	160	167
16	57	62	163	172
17	58	65	163	174
18	58	68	163	178

Data and Observations: Students will observe that females have an earlier growth spurt than males, but on the average, males grow taller and heavier than females.

Name Date Class

INVESTIGATE BioLab Average Growth Rate in Humans, *continued*

Chapter 35

Analyze and Conclude

1. **Analyzing Data** During what ages do females and males increase the most in mass? In height?
   Mass: females, ages 11–13; males, ages 12–14. Height: females, ages 9–14; males, ages 12–15
2. **Analyzing Data** Interpret the data to find if the average growth rate is the same in males and females.
   Average growth is the same until puberty. At puberty, females have an earlier growth spurt than males, but on the average, males grow taller and have greater mass than females.
3. **Thinking Critically** How can you explain the differences in growth rates between males and females?
   Timing of puberty and the production of sex hormones differ between males and females. For example, males produce more of the hormone testosterone than females. Testosterone increases growth in muscle and bone mass in males.
4. **Relating Concepts** Why do you think male and female growth rates increase during the teen years?
   The reproductive hormones released during the teen years increase the growth rate.

Name Date Class

MiniLab 36-1 Distractions and Reaction Time

Experimenting

Have your ever tried to read while someone is talking to you? What effect does such a distracting stimulus have on your reaction time?

Procedure

1. Work with a partner. Sit facing your partner as he or she stands.
2. Have your partner hold the top of a meterstick above your hand. Hold your thumb and index finger about 2.5 cm away from either side of the lower end of the meterstick without touching it.
3. Tell your partner to drop the meterstick straight down between your fingers.
4. Catch the meterstick between your thumb and finger as soon as it begins to fall. Measure how far it falls before you catch it. Practice several times.
5. Run ten trials, recording the number of centimeters the meterstick drops each time. Average the results.
6. Repeat the experiment, this time counting backwards from 100 by fives (100, 95, 90, . . .) as you wait for your partner to release the meterstick.

Analysis

1. Did your reaction time improve with practice? Explain.
As students learn to anticipate the drop, their reaction time should improve.

2. How was your reaction time affected by the distraction (counting backward)?
The distraction probably increased their reaction time.

3. What other factors, besides distractions, would increase reaction time?
Answers may include being tired, sick, hungry, or otherwise preoccupied.

Name Date Class

MiniLab 36-2 Interpret a Drug Label

Analyzing Information

One common misuse of drugs is not following the instructions that accompany them. Over-the-counter medicines can be harmful—even fatal—if they are not used as directed. The Food and Drug Administration requires that certain information about a drug be provided on its label to help the consumer use the medicine properly and safely.

Procedure

1. The photograph on page 991 of your text shows a label from an over-the-counter drug. Read it carefully.
2. Use the data table below. Fill in the table using information on the label.

Information from a drug label

People with these conditions should avoid this drug	Possible side effects	This drug should not be taken with these medicines	Symptoms this drug will relieve	Correct dosage

Analysis

1. What is a side effect? What side effects are caused by this drug?
A side effect is any effect of the drug other than the one it is designed to produce. Side effects of this drug include drowsiness.

2. Why should a person never take more than the recommended dosage?
The recommended dosage is the one that has been tested and proven to be safe. Over-the-counter medicines are potentially harmful if not used as directed.

3. How are over-the-counter drugs different from prescription drugs?
Over-the-counter drugs are available without a doctor's prescription. They are not as strong as prescription medicines.

Expected Results: Student tables should show the following information: among others, children under 12 years, people with high blood pressure, heart disease, diabetes, thyroid disease, asthma, or emphysema should avoid this drug; drowsiness is a possible side effect; should not be taken with antihypertensive or antidepressant drugs; will relieve nasal congestion associated with the common cold, hay fever, or sinusitis; the correct dosage is one tablet every 12 hours.

Name Date Class

DESIGN YOUR OWN BioLab

What drugs affect the heart rate of *Daphnia*?

Chapter 36

Preparation

Problem

What legally available drugs are stimulants to the heart? What legal drugs are depressants? Because these drugs are legally available, are they less dangerous?

Hypotheses

Based on what you learned in this chapter, which of the drugs listed under Possible Materials do you think are stimulants? Which are depressants? How will they affect the heart rate in *Daphnia*? Make a hypothesis concerning how each of the drugs listed will affect heart rate.

Objectives

In this BioLab, you will:

- **Measure** the resting heart rate in *Daphnia.*
- **Compare** the resting heart rate with the heart rate when a drug is applied.

Possible Materials

aged tap water
Daphnia culture
dilute solutions of coffee, tea, cola, ethyl alcohol, tobacco, and cough medicine (destromethorphen)
dropper
microscope
microscope slide

Safety Precautions

Do not drink any of the solutions used in this lab. Always wear goggles in the lab. Use caution when working with a microscope, microscope slides, and glassware.

Skill Handbook

Use the **Skill Handbook** if you need additional help with this lab.

Plan the Experiment

1. Using a dropper, place a single *Daphnia* crustacean on a slide.
2. Observe the animal on low power and find its heart.
3. Design an experiment to measure the effect on heart rate of four of the drug-containing substances in the Possible Materials list.
4. Design and construct a data table for recording your data.

Check the Plan

1. Be sure to consider what you will use as a control.
2. Plan to add two drops of a drug-containing substance directly to the slide.
3. When you are finished testing one drug, you will need to flush the used *Daphnia* with the solution into a beaker of aged tap water provided by your teacher. Plan to use a new *Daphnia* for each substance tested.
4. ***Make sure your teacher has approved your experimental plan before you proceed further.***
5. Carry out your experiment. CAUTION: ***Wash your hands with soap and water immediately after making observations.***

Name Date Class

DESIGN YOUR OWN BioLab

What drugs affect the heart rate of *Daphnia*?, *continued*

Chapter 36

Analyze and Conclude

1. **Making Inferences** Which drugs are stimulants? Which are depressants?
   Stimulants are coffee, tea, cola, and tobacco. Cough medicine may also be listed. Depressants are ethyl alcohol and cough medicine if it contains dextromethorphen hydrobromide.
2. **Checking Your Hypotheses** Compare your predicted results with the experimental data. Explain whether or not your data support your hypotheses regarding the drugs' effects.
   Some students' hypotheses will be confirmed by their data; others' will be rejected.
3. **Drawing Conclusions** How do the drugs affect the heart rate of this animal?
   Stimulants speed up the animal's heart rate. Depressants slow the heart rate.
4. **Analyzing the Procedure** How would you alter your experiment if you did it again?
   Answers will vary.

Data and Observations: Sample Data

Drug	Heart rate/min
No drug	240
Coffee	270
Cola	270
Tea	260
Ethyl alcohol	215
Tobacco	300
Cough medicine	heart rate varies with brand

Name Date Class

MiniLab 37-1 Checking Your Pulse

Experimenting

The heart speeds up when the blood volume reaching your right atrium increases. It also speeds up when the level of carbon dioxide in the blood rises. The number of heartbeats per minute is your heart rate, which can be measured by taking your pulse.

Procedure

1. Use the data table below.
2. Have a classmate take your resting pulse for 60 seconds while you are sitting at your lab table or desk. Use the photo on page 1013 of your text as a guide to finding your radial pulse.
3. Record your pulse in the table.
4. Repeat steps 2 and 3 four more times, then calculate your average resting pulse rate. Switch roles and take your classmate's resting pulse.
5. Exercise by doing "jumping jacks" for one minute.
6. Have your classmate take your pulse for 60 seconds immediately after exercising and record the value in the data table.
7. Repeat steps 5 and 6 four more times. Switch roles again with your classmate.

Data Table

Heart rate (beats per minute)		
Trial	**Resting**	**After exercise**
1		
2		
3		
4		
5		
Total		
Average		

Expected Results: Resting pulses will be around 80–90 beats per minute. After exercise, beats will be close to 120 per minute.

Name Date Class

MiniLab 37-1 Checking Your Pulse, *continued*

Analysis

1. Explain why your pulse is a means of indirectly measuring heart rate.
   A surge of blood through arteries corresponds with each ejection of blood from the heart.
2. Use actual values from your data table to describe the changes that occur to your heart rate when exercising.
   Answers will vary. Average pulse after exercise is higher than resting pulse.
3. Suppose the amount of blood pumped by your left ventricle each time it contracts is 70 mL. Calculate your cardiac output (70 mL × heart rate per minute) while at rest and just after exercise.
   70 mL × average resting pulse per minute = resting cardiac output in mL per minute
   70 mL × average exercising pulse per minute = exercising cardiac output in mL per minute

Name Date Class

MiniLab 37-2 Testing Urine for Glucose

Experimenting

Glucose is a sugar that is needed by the body and is normally not present in the urine. When the concentration of glucose becomes too high in the blood, as happens with diabetes, glucose is filtered out by the kidneys.

Procedure

1. Use the data table.
2. Using a grease pencil, draw two circles on a glass slide. Mark one circle N, the other A.
3. Use a clean dropper to add two drops of "normal urine" to the circle marked N.
4. Use a clean dropper to add two drops of "abnormal urine" to the circle marked A.
5. Hold a small strip of glucose test paper in a forceps and touch it to the liquid in the drop labeled N. Remove it, wait 30 seconds, and record the color. A green color means glucose is present.
6. Use a new strip of glucose test paper to test drop A and record the color.
7. Test several unknown "urine" samples for the presence of glucose. Use a clean slide for each test.

Data Table

Urine sample	Color of test paper	Glucose present?
Normal (N)		
Abnormal (A)		
Unknown X		
Unknown Y		
Unknown Z		

Expected Results: Test paper turns green in the presence of glucose. Water causes no color change.

Analysis

1. Which of the "unknown" samples could be from a person who has diabetes?
Any unknown that tests positive for glucose could indicate diabetes. High sugar intake may also result in glucose elimination in urine.
2. Which part of the test procedure could be considered your control? Explain your answer.
testing of normal and abnormal urine before testing of unknowns
3. How could you explain your results if a test of normal urine indicated the presence of glucose?
Possible answers: error in labeling, contamination of samples, normal urine was mixed with abnormal urine by accident, not using clean droppers

Name Date Class

Chapter 37

INVESTIGATE BioLab

Measuring Respiration

Preparation

Problem
How can you measure respiratory rate and estimate tidal volume?

Objectives
In this BioLab, you will:
- **Measure** resting breathing rate.
- **Estimate** tidal volume by exhaling into a balloon.
- **Calculate** the amount of air inhaled per minute.

Materials
round balloon
string (1 m)
metric ruler
clock or watch with second hand

Skill Handbook
Use the **Skill Handbook** if you need additional help with this lab.

Procedure

Part A: Breathing Rate at Rest
1. Use Data Table 1.
2. Have your partner count the number of times you inhale in 30s.
3. Repeat step 2 two more times.
4. Calculate the average number of breaths.
5. Multiply the average number of breaths by two to get the average resting breathing rate in breaths per minute.

Data Table 1

Resting breathing rate	
Trial	**Inhalations in 30 sec**
1	
2	
3	
Average number breaths	
Breaths/min	

Part B: Estimating Tidal Volume
1. Use Data Table 2.
2. Take a regular breath and exhale normally into the balloon. Pinch the balloon closed.
3. Have a partner fit the string around the balloon at the widest part.
4. Measure the length of the string, in centimeters, around the circumference of the balloon. Record this measurement.
5. Repeat steps 2–4 four more times.
6. Calculate the average circumference of the five measurements.
7. Calculate the average radius of the balloon by dividing the average circumference by 6.28 (which is approximately equal to 2π).
8. Tidal volume is the amount of air expelled during a normal breath. Tidal volume can be determined using the balloon radius and the formula for determining the volume of a sphere.

$$\text{Volume} = \frac{4\pi r^3}{3}$$

where r = radius and $\pi = 3.14$. Calculate the average tidal volume using the average balloon radius.

Name Date Class

INTERNET BioLab

Measuring Respiration, *continued*

Chapter 37

Data Table 2

Tidal volume	
Trial	**String measurement**
1	
2	
3	
4	
5	
Avg. circumference	
Avg. radius	
Avg. tidal volume	

9. Your calculated volume will be in cubic centimeters: $1\ cm^3 = 1$ mL.

Part C: Amount of Air Inhaled
1. Use Data Table 3.
2. Multiply the average tidal volume by the average number of breaths per minute to calculate the amount of air you inhale per minute.
3. Divide the number of milliliters of air by 1000 to get the number of liters of air you inhale per minute.

Data Table 3

Amount of air inhaled	
mL/min	
L/min	

Analyze and Conclude

1. **Making Comparisons** Compare your average number of breaths per minute and tidal volume per minute with those of other students.
Average breaths per minute and tidal volume per minute will differ among students.

2. **Thinking Critically** An average adult inhales 6000 mL of air per minute. Compare your estimated average volume of air with this figure. What factors could account for any differences?
Answers may vary from the average due to age, sex, size, and athletic condition.

3. **Making Predictions** Predict what would happen to your resting breathing rate after exercise.
After exercising, the breathing rate will be higher.

Data and Observations: Answers among students could vary greatly. Average breathing rate is 11 to 12 breaths per minute. Tidal volume should be approximately 280 mL. The amount of air inhaled should be 5 to 6 L per minute.

Name Date Class

MiniLab 38-1 Examining Sperm and Egg Attraction

Observing and Inferring

Most animals that reproduce by external fertilization live in water and release their sperm and eggs into the water. Somehow, a sperm and egg must meet. One adaptation of these animals that helps ensure fertilization is the release of thousands of sex cells at one time. The large number of sex cells increases the odds that at least some eggs will be fertilized. Chemical attraction could be another adaptation that encourages fertilization. If eggs give off a chemical that attracts sperm, each sperm would have help "finding" an egg.

Procedure

1. Place a dropperful of sea urchin eggs on a microscope slide. **CAUTION:** ***Use care when working with a microscope and microscope slides.***
2. While observing eggs under the microscope, add a drop of sea urchin sperm to the eggs.

Analysis

1. Describe the motion of a single sperm.
   The sperm swim like tadpoles with tails whipping back and forth.
2. What cell structures are involved in providing energy for the sperm motion?
   mitochondria
3. Are the sperm attracted to the eggs? How do you know?
   Yes, they swim toward and gather around the egg.

Expected Results: Sperm will collect around the egg, indicating that they are attracted to the egg.

Name Date Class

MiniLab 38-2 Making a Graph of Fetal Size

Making and Using Graphs

You started out as a single cell. That cell divided by the process of mitosis to produce organ systems capable of maintaining an independent existence outside your mother's uterus. During the time you were in your mother's uterus, major changes took place. One of these changes involved your growth in length.

Table 38.1: Growth of a Fetus

Source of sample	Time after fertilization	Size
First trimester	3 weeks	3 mm
	4 weeks	6 mm
	6 weeks	12 mm
	7 weeks	2 cm
	8 weeks	4 cm
	9 weeks	5 cm
	3 months	7.5 cm
Second trimester	4 months	15 cm
	5 months	25 cm
	6 months	30 cm
Third trimester	7 months	35 cm
	8 months	40 cm
	9 months	51 cm

Procedure

1. Prepare a graph that plots time on the horizontal axis and length in centimeters on the vertical axis. Equally divide the horizontal axis into nine months. Then equally divide each of the first three months into four weeks.
2. Plot the data in *Table 38.1* on your graph.

Expected Results: Graphs will show that the embryo grows fastest during the early periods of development.

Analysis

1. When is the fastest period of growth?
   The embryo doubles in size during two 1-week periods: 3 to 4 weeks and 7 to 8 weeks.
2. What structures are developing during this period of growth?
   All body systems are beginning to form.
3. At what point does growth begin to slow down?
   5th month

Name Date Class

Investigate BioLab

What hormone is produced by an embryo?

Chapter 38

Preparation

Problem
How can you test for the presence of hCG?

Objectives
In this BioLab, you will:
- **Model** the chemicals used to test for the presence of hCG.
- **Interpret** the results of chemical reactions involving hCG in a pregnant and nonpregnant female.

Materials
scissors
heavy paper
tracing paper

Safety Precautions
Handle scissors with caution.

Skill Handbook
Use the **Skill Handbook** if you need additional help with this lab.

Procedure

1. Use the data table below.
2. Copy models **A**, **B**, and **C** on page 1048 of your text onto tracing paper.
3. Copy the tracings onto heavy paper and cut them out. You will need 4 models of **A**, 4 models of **B**, and 1 model of **C**.
4. Model **A** represents a molecule of the hCG hormone. Model **B** represents a chemical called anti-hCG hormone. Model **C** represents a chemical that has four hCG molecules attached to it.
5. Note that the shapes of hCG and anti-hCG join together like puzzle pieces. These two chemicals react, or join together, when both are present in a solution. The shapes of anti-hCG and Chemical C also join, indicating that they chemically react when both are present. The combination of Chemical C and anti-hCG is green. Chemical C without anti-hCG attached is colorless.
6. Model the following events for the "Not pregnant" condition. Record them in the data table using drawings of the models.
   a. The hormone hCG is not present in the urine.
   b. Anti-hCG is added to a urine sample, then chemical C is added.
   c. Draw the resulting chemical in the data table and indicate the color that appears.

Data Table

Condition	hCG in urine?	+ Anti-hCG	= Joined hCG and anti-hCG?	+ Chemical C with anti-hCG?	Color
Not pregnant	———		———		Green
Pregnant					Colorless

Name Date Class

Investigate BioLab

What hormone is produced by an embryo?, *continued*

Chapter 38

7. Model the following events for the "Pregnant" condition. Record them in the data table using drawings of the models.
   a. The hormone hCG is present.
   b. Anti-hCG is added to urine, then Chemical C is added.
   c. Draw the resulting chemical in the data table and indicate the color that appears.

Analyze and Conclude

1. **Analyzing** Explain the origin of hCG in a pregnant female.
   The chorion begins to release hCG as early as eight days after fertilization.
2. **Analyzing** Explain why hCG is absent in a nonpregnant female.
   There is no chorion to release hCG.
3. **Concluding** Describe the roles of anti-hCG and Chemical C in both tests.
   If pregnant, anti-hCG joins with hCG, preventing attachment to chemical C's hCG. No color appears. If not pregnant, anti-hCG attaches to chemical C's hCG and color appears.
4. **Observing and Inferring** Explain why anti-hCG is added to the sample before Chemical C is added.
   To get an accurate result, hCG and anti-hCG must have a chance to bind before Chemical C is added.

Name Date Class

MiniLab 39-1 Testing How Diseases are Spread

Experimenting

Microorganisms cannot travel over long distances by themselves. Unless they are somehow transferred from one animal or plant to another, infections will not spread. One method of transmission is by direct contact with an infected animal or plant.

Procedure

1. Label four plastic bags 1 to 4.
2. Put a fresh apple in bag 1 and seal the bag.
3. Rub a rotting apple over the entire surface of the remaining three apples. The rotting apple is your source of pathogens. **CAUTION:** ***Make sure to wash your hands after handling the rotting apple.***
4. Put one of the apples in bag 2.
5. Drop one apple to the floor from a height of about 2 m. Put this apple in bag 3.
6. Use a cotton ball to spread alcohol over the last apple. Let the apple air-dry and then place it in bag 4.
7. Store all of the bags in a dark place for one week.
8. Compare the apples and record your observations. **CAUTION:** ***Give all apples to your teacher for proper disposal.***

Analysis

1. What was the purpose of the fresh apple in bag 1?
   Apple 1 is a control for comparison.
2. Explain what happened to the rest of the apples.
   Apple 2 decayed the most, Apple 3 had brown spots, and Apple 4 may have shown little or no change.
3. Why is it important to clean a wound with disinfectant?
   Cleaning a wound with alcohol may destroy some pathogens and help prevent infection.

Expected Results: Apples 2 and 3 will develop brown spots and decay. Apple 1 will show little or no change. Apple 4 will show little or no change, because the alcohol may have inhibited the decay.

Name Date Class

MiniLab 39-2 Distinguishing Types of White Blood Cells

Observing and Inferring

The human immune system includes five types of white blood cells found in the bloodstream: basophils, neutrophils, monocytes, eosinophils, and lymphocytes.

Procedure

1. Use the data table below.
2. Mount a prepared slide of blood cells on the microscope and focus on low power. Turn to high power and look for white blood cells. **CAUTION:** ***Use caution when working with microscope slides.***
3. Find a neutrophil, monocyte, eosinophil, and lymphocyte. You may see a basophil, although they are rare. Refer to the *Inside Story* for photos of these cells.
4. Count a total of 50 white blood cells, and record how many of each type you see.
5. Calculate the percentage by multiplying the number of each cell type by two. Record the percentages. Diagram each cell type.

Data Table

Type of white blood cell	Number counted	Percent	Diagram
Neutrophil			
Monocyte			
Basophil			
Lymphocyte			
Eosinophil			

Expected Results: The common percentages of white blood cells are: neutrophils, 60–70%; lymphocytes, 20–25%; monocytes, 3–8%; eosinophils, 2–4%; and basophils, 0.5–1%.

Analysis

1. Which type of white blood cell was most common? Second most common?
   neutrophils, lymphocytes
2. How do red and white blood cells differ?
   White blood cells have a nucleus; red blood cells don't.

Name Date Class

INTERNET BioLab

Getting On-line for Information on Diseases

Chapter 39

Preparation

Problem

How can you use the Internet to obtain current research information on different diseases?

Objectives

In this BioLab, you will

- **Choose** five communicable and five noncommunicable diseases for study.
- **Use the Internet** to gather information.
- **Collect data** on the ten diseases and record it in a table.

Materials

access to the Internet

Skill Handbook

Use the **Skill Handbook** if you need additional help with this lab.

Procedure

1. Use the two data tables.
2. Choose five communicable and five noncommunicable diseases you wish to investigate.
3. List the diseases in your data tables. Try to make your disease choices as specific as possible. For example, cancer as a topic is too broad. Instead, limit your choice to a specific type of cancer, such as breast cancer, prostate cancer, or Hodgkin's disease.
4. Go to the Glencoe Science Web Site at **www.glencoe.com/sec/science** to find helpful links to research information for this BioLab.
5. Be sure to complete the last two rows asking for current research findings and your sources of information.

Data Table 1

Communicable diseases	1	2	3	4	5
Disease name					
Organism responsible					
Classification of organism					
Mode of transmission					
Symptoms					
Treatment					
Current research					
Source of information					

Data and Observations: Student data and observations will vary, depending on the diseases selected and the information resources used.

Name Date Class

INTERNET BioLab

Getting On-line for Information on Diseases, *continued*

Chapter 39

Data Table 2

Noncommunicable diseases	1	2	3	4	5
Disease name					
Symptoms					
Organs affected					
Age group affected					
Treatment					
Current research					
Source of information					

Analyze and Conclude

1. **Defining** What is a pathogen? Provide several examples.
   A pathogen is a disease-producing agent. Pathogens include bacteria, viruses, parasites, and fungi.
2. **Comparing** Describe the difference between a communicable and a noncommunicable disease. Provide several examples of each.
   Communicable diseases can be passed from one person to another, while noncommunicable diseases cannot.
3. **Thinking Critically** What are vectors? Are they associated with communicable or noncommunicable diseases? Explain your answer.
   Vectors are carriers of a pathogen. Since noncommunicable diseases are not caused by pathogens, there are no vectors involved in their transmission.
4. **Applying Concepts** Explain why the table for noncommunicable diseases does not have a column for organism responsible or method of transmission.
   Noncommunicable diseases are not caused by an organism or pathogen.
5. **Using the Internet** What is one advantage of getting information on disease research by way of the Internet rather than from textbooks or an encyclopedia?
   Internet information can be updated more frequently than information printed in textbooks and encyclopedias.